Theory and Best Practices in Science Communication Training

This edited volume reports on the growing body of research in science communication training and identifies best practices for communication training programs around the world.

Theory and Best Practices in Science Communication Training provides a critical overview of this emerging field. It analyzes the role of communication training in supporting scientists' communication and engagement goals, including their motivations to engage in training, the design of training programs, methods for evaluation, and frameworks to support the role of communication training in helping scientists reach their goals. Overall, this collection reflects on the growth of the field and provides direction for developing future researcher–practitioner collaborations.

With contributions from researchers and practitioners from around the world, this book will be of great interest to students, scholars and professionals within this emerging field.

Todd P. Newman is an assistant professor in the Department of Life Sciences Communication at the University of Wisconsin-Madison, USA where he teaches courses on science communication, strategic communication, and marketing. Newman is the co-author of *Brand* (2018) – which examines the role of brand strategy in society, including scientific debates – and previously conducted research on science communication training as a postdoctoral associate at the University of Connecticut and the Alan Alda Center for Communicating Science at Stony Brook University.

Routledge Studies in Environmental Communication and Media

The Discourses of Environmental Collapse
Imagining the End
Edited by Alison E. Vogelaar, Brack W. Hale and Alexandra Peat

Environmental Management of the Media
Policy, Industry, Practice
Pietari Kääpä

Participatory Networks and the Environment
The BGreen Project in the US and Bangladesh
Fadia Hasan

Participatory Media in Environmental Communication
Engaging Communities in the Periphery
Usha Harris

Journalism, Politics, and the Dakota Access Pipeline
Standing Rock and the Framing of Injustice
Ellen Moore

Environmental Literacy and New Digital Audiences
Patrick Brereton

Reporting Climate Change in the Global North and South
Journalism in Australia and Bangladesh
Jahnnabi Das

Theory and Best Practices in Science Communication Training
Edited by Todd P. Newman

For more information about this series, please visit: www.routledge.com/Routledge-Studies-in-Environmental-Communication-and-Media/book-series/RSECM

Theory and Best Practices in Science Communication Training

Edited by Todd P. Newman

Routledge
Taylor & Francis Group
LONDON AND NEW YORK

earthscan
from Routledge

First published 2020
by Routledge
2 Park Square, Milton Park, Abingdon, Oxon OX14 4RN

and by Routledge
52 Vanderbilt Avenue, New York, NY 10017

Routledge is an imprint of the Taylor & Francis Group, an informa business

British Library Cataloguing-in-Publication Data
A catalogue record for this book is available from the British Library

Library of Congress Cataloging-in-Publication Data
A catalog record has been requested for this book

ISBN: 978-1-138-47815-2 (hbk)
ISBN: 978-1-351-06936-6 (ebk)

Typeset in Times New Roman
by Wearset Ltd, Boldon, Tyne and Wear

Contents

Illustrations

Figures

Tables

Contributors

Fred Balvert, MSc, is Science Communicator and Head of the Congress agency at Erasmus MC University Medical Center Rotterdam, The Netherlands. He is Senior Lecturer in Science Communication Management at Rhein-Waal University of Applied Sciences, Germany, and Media Advisor of the Erice International School of Science Journalism, Italy. He studied Leisure Management (BA) and Public Administration (MSc). He is one of the authors of the handbook 'Prepare for 15 minutes of fame: Media contacts for researchers.'

Ayelet Baram-Tsabari, PhD, is associate professor at the Faculty of Education in Science and Technology at the Technion – Israel Institute of Technology and heads the "Applied Science Communication" research group. Prof. Baram-Tsabari founded the Israeli Science Communication Conference series and is an elected member of the scientific committee of the Public Communication of Science and Technology Network (PCST), which aims to improve science communication worldwide. She is an elected member of the Israel Young Academy and served as Chairwoman of its Communication Committee. Baram-Tsabari is a member of the Advisory Board for the US National Academy of Sciences' LabX public engagement program, and a member of the Israeli Centers of Research Excellence (I-CORE) on "Learning in a Networked Society" and "Taking Citizen Science to School," among other professional activities. Baram-Tsabari's research focuses on bridging science education and science communication scholarship, identifying people's interests in science, building on people's authentic interests in science to teach and communicate science in more meaningful and personally relevant ways, and supporting scientists in becoming effective science communicators. She has published extensively in the leading journals in both science education and science communication.

Yael Barel-Ben David is a PhD student at the Faculty of Education in Science and Technology at the Technion – Israel Institute of Technology and part of the "Applied Science Communication" research group. She is a graduate fellow of the Israeli Center of Research Excellence (I-CORE) on "Learning in a Networked Society." Barel-Ben David holds a BSc in Biology and

Philosophy from the Hebrew University in Jerusalem (2010), and MSc (2015, magna cum laude) in science communication. Barel-Ben David's practical science communication experience includes guiding, supervising and developing programs at the Bloomfield Science Museum in Jerusalem, repeatedly participating in a TV science show, and national finalist of the FameLab competition. She leads science communication workshops and is regularly invited to lecture in teachers' professional development and undergraduate courses, as well as International workshops. Barel-Ben David's research focuses on science communication training programs for scientists – their impact, effectiveness, and usefulness for participants; goals and motivations of practitioners, and public's interaction with training outcomes.

Nichole Bennett, MA, is a graduate student researching science communication in the PhD program at The Stan Richards School of Advertising and Public Relations at The University of Texas at Austin. Her work focuses on the processes underlying scientists' engagement with the public and incorporating strategic communication into training programs to support these scientists.

John C. Besley, PhD, is the Ellis N. Brandt Professor of Public Relations at Michigan State University. He studies how views about decision processes affect perceptions of science and technology with potential health or environmental impacts. This work emphasizes the need to look at both citizens' perceptions of decision makers and decision makers' perceptions of the public.

Dominique Brossard, PhD, is a professor and the Chair of the Department of Life Sciences Communication at the University of Wisconsin-Madison, USA. Brossard's research agenda focuses on the intersection between science, media and policy. She is an internationally known expert in public opinion dynamics related to controversial scientific issues. She has published more than 100 research articles in outlets such as *Science* and *Public Understanding of Science*, and has been an expert panelist for the National Academy of Sciences on various occasions. Brossard earned her MS in plant biotechnology from the Ecole Nationale d'Agronomie de Toulouse (France) and PhD in communication from Cornell University, USA

Anthony Dudo, PhD, is an associate professor in The Stan Richards School of Advertising and Public Relations at The University of Texas at Austin. He studies scientists' public engagement activities, media representations of science and environmental issues, and public perceptions of science. He is also the program director for science communication at the UT Center for Media Engagement.

Declan Fahy, PhD, is an assistant professor at the School of Communications, Dublin City University, Ireland. He is the author of *The New Celebrity Scientists: Out of the Lab and Into the Limelight* (2015). Between 2006 and 2008, he was coordinator of the European Science Communication Network (ESConet).

Toss Gascoigne is a visiting Fellow at the Centre for Public Awareness of Science at the Australian National University, in Canberra, Australia. He has run hundreds of training workshops in Australia, the Pacific and a dozen other countries, working with Jenni Metcalfe. These have focused on science communication, how to reach different audiences and explain complex concepts. He is a member of the international Network for the Public Understanding of Science and Technology, and helped transform PCST into a body with international reach. He is a founding member of Australian Science Communicators which he helped to establish in 1994. Toss is a former President and life member of both Australian Science Communicators and the PCST Network. He is interested in the interface between science and policy, often using the media as a tool, and for 15 years headed national organizations in Australia. He worked as Executive Director of the Federation of Australian Scientific and Technological Societies (FASTS) and the Council for the Humanities, Arts and Social Sciences (CHASS). He has published on the history of modern science communication, on whether the field could be considered a discipline, and science advocacy. He has also written on the establishment of 'Science meets Parliament', a successful Australian initiative which allows scientists to meet national politicians to make the case for science and research to the government.

Emily Howell is a doctoral candidate in the Nelson Institute for Environmental Studies at the University of Wisconsin-Madison and works with the Department of Life Sciences Communication, researching how to communicate controversial and policy-relevant science issues, such as fracking, human genome editing, and synthetic biology. Her work focuses on the role of values in shaping reasoning and opinions toward potentially-polarizing science issues to better understand how to facilitate effective science communication and public engagement with science. She has her BA in History of Science and Earth & Planetary Sciences from Harvard and her MS in Environment & Resources and Energy Analysis and Policy.

Louise Kuchel, PhD, GCHEd, SFHEA, is a biology lecturer and science education researcher at the University of Queensland, Australia. She trained as a scientist and worked as a biologist in industry and research for 15 years before transitioning to teaching and education research in 2008; she has always loved working at the interface between disciplines. Her research focuses on teaching and learning of science communication by scientists, in higher education and environmental and conservation outreach. She currently teaches biology, science communication and quantitative skills in science at Bachelor and Masters levels, running occasional workshops for PhD students and postdoctoral fellows. Louise has been recognized as a Senior Fellow of the Higher Education Academy.

Nicole J. Leavey, PhD, is a Postdoctoral Research Associate at the Alan Alda Center for Communicating Science. She holds a PhD in Technology, Policy and Innovation and an MS in Technology Systems Management from Stony Brook University, College of Engineering and Applied Sciences. Her research interests focus on public engagement actions of scientists, science communication training and gender roles. She also contributes to evidence-based curriculum design and delivers communication training for scientists at both academic and professional institutions.

Brenda L. MacArthur, PhD, is a Health Communication Postdoctoral Associate at the Alan Alda Center for Communicating Science® at Stony Brook University. Dr. MacArthur's research is centered on the intersection of health and instructional communication to improve the quality and continuity of health care through evidence-based communication training. Her line of research also explores topics such as health literacy, organizational socialization, interdisciplinary teams, patient adherence, and communication motives. Dr. MacArthur holds a PhD in Health Communication from George Mason University, an MA in Communication Studies from Texas State University, and a BA in Communication from Bryant University.

Jenni Metcalfe is the Founder and Director of Econnect Communication, established in 1995 to help scientists communicate about their research. She has been a science communicator for almost 30 years, working as a journalist, practitioner, university lecturer and researcher. Jenni has published many papers and articles on science communication. Jenni has been a member of the scientific committee of the International Public Communication of Science and Technology (PCST) Network since 1996. She is currently Chair of the PCST's Conference Program Committee with the next conference being held in Aberdeen Scotland in 2020. Jenni is completing her PhD in science communication at the Australian National University. Jenni is passionate about doing science communication that creates a positive difference to people's lives. She believes that every person has a right to understand and engage with science so they can make more informed decisions about issues and opportunities that affect their lives.

Amanda E. Ng, MPH, is a doctoral student of Epidemiology at the University of Maryland, College Park. She received her Master's in Public Health from the Renaissance School of Medicine at Stony Brook University. She is interested in investigating the social determinants of health, primarily childhood adversity, and how they may be related to the development of chronic disease in adulthood. As a Research Associate for the Alan Alda Center for Communicating Science, Ms. Ng is also passionate about increasing the public's understanding of science through audience-focused communication.

Tzipora Rakedzon, PhD, serves as the coordinator and a lecturer of Graduate Academic Writing in the Department of Humanities and Arts at the Technion – Israel Institute of Technology. She has been teaching a variety of writing and communication courses for over 20 years. Tzipora received her PhD at the Technion in science communication at the Department of Education in

Technology and Science, and her BA and MA in linguistics from Haifa University. She is also a lecturer and the coordinator of the English Program at the Guangdong Technion – Israel Institute of Technology in China. As well, Tzipora oversees the new writing center which is about to open at the Technion. Her research interests include teaching and assessing scientific and professional communication, especially writing and vocabulary.

Brooke Smith is the Director of Public Engagement with Science at The Kavli Foundation, where she works to strengthen the field of science communication and public engagement with science. She was the first Executive Director of COMPASS, where she helped build the organization to be a leading science communication training and boundary organization. She has led science communication and engagement activities at universities and non-profits, as well as policy development and implementation with federal and state agencies. She is passionate about empowering scientists to be effective in public discourses. She has an undergraduate degree from Duke University, and a Master's of science in oceanography from Oregon State University.

Shupei Yuan, PhD, is an assistant professor in the Department of Communication at Northern Illinois University. Her research focuses on the support and factors that influence scientists' engagement with the public, and the strategic communication styles in the context of risk, science, and health communication.

Introduction

Todd P. Newman

At the 2007 annual meeting of the American Association for the Advancement of Science – the largest general scientific society in the world – Google co-founder Larry Page delivered the keynote lecture with a message to the scientists in the room: "Science has a really serious marketing problem and nobody pays attention to that since none of the marketers work for science." In other words, if scientists really want their work to reach broader audiences, scientists and the various actors at the science–society interface, such as governments, universities, funders, and professional societies, need to understand the importance of communication.

Efforts by the scientific community to "educate" the public on general science or science-specific issues overall tend to fall short of desired outcomes, leading many leaders from the scientific community to recognize the need for public engagement and communication (Cicerone, 2010; Leshner, 2003, 2015). Given the lack of connectivity between scientists and society, scientists must develop closer ties to different publics and engage in bidirectional communication. This type of communication reflects the need for science and scientists to integrate the many different needs and values that science meets for society.

The push for this effort rests on a strong foundation: Internationally, the public has a great deal of trust in scientists and views the benefits of science to outweigh the risks (National Science Board, 2018). It is within this context that research within the "science of science communication" seeks to understand the various factors that contribute to how the public thinks about scientific issues as well as frameworks for effective engagement between scientists and different audiences (see Hall-Jamieson, Kahan, & Scheufele, 2017). The research within this field led to the recent (2017) publication by the National Academies of Sciences, Medicine, and Engineering, "Communicating Science Effectively: A Research Agenda", which provided a framework for the field of science communication moving forward.

For scientists, the need for effective communication has a number of practical applications. For instance, scientists across the STEM (Science, Technology, Engineering, and Mathematics) fields need to explain the importance and application of their research to policymakers and commercial partners in order to yield benefits to society. Likewise, scientists need to describe the results of their research to journalists to keep the public informed. Moreover, several federal research-funding bodies, including those in the US, UK, and European Union, have included a "broader impacts" or "dissemination" clause to make sure that the benefits of federally funded research to society are communicated.

Despite these changing norms, many scientists do not receive formal training in communication or the social sciences. Yet many scientists indicate high levels of willingness to engage and view engagement to be beneficial overall. Therefore, the need for effective science communication has led to the development of science communication training courses integrated into graduate STEM education, as well as a growing number of research centers and organizations providing communication training to university and government-based scientists, scientists in private industry, as well as medical professionals. This training consists of various activities, including courses, workshops, and seminars that rely on a variety of techniques to improve communication effectiveness.

With the growth of science communication training programs and courses around the world, there is a growing community of researchers focused on understanding the role of communication training in supporting scientists' communication efforts as well as scientists' motivation to seek training. More recently, researchers and practitioners recognize the need to develop frameworks for the evaluation of science communication training programs, including ways to measure communication effectiveness.

Yet the research on science communication training remains scattered across various discipline-specific journals in education, communication, and journalism. Recent edited volumes in science education and informal learning (Patrick, 2017; Van der Sanden & De Vries, 2016) as well as dedicated special issues, such as the *International Journal of Science Education, Part B* have focused on these topics. However, to date there is no comprehensive volume of the different strands of research that examine trends in science communication training which may serve as a reference of best practices as well as an overview of the field for science communication researchers, practitioners, and trainers as well as other interested readers.

The purpose of this edited volume is to bring together some of the leading researchers and practitioners from around the world in the field of science communication training. Bridging these different perspectives will provide a broad overview of current trends in science communication training, including best practices and areas in need of future research. *Theory and Best Practices in Science Communication Training* contains 10 original chapters broken down into three parts.

Part I, "The Scientist as a Strategic Communicator," contains four chapters examining the role of the scientist as a strategic communicator and the extent to which communication training supports these efforts.

In Chapter 1, Nichole Bennett, Anthony Dudo, Shupei Yuan, and John Besley provide an overview of their research program focused on the role of communication strategy for scientists as well as the trainers who support these efforts. The authors provide a comprehensive overview of the public engagement scholarship, including the frequency of scientists' communication activity, the factors underlying their willingness to communicate, and how they approach communication. The authors find that despite their efforts, scientists are not strategic communicators and prioritize objectives that are unlikely to change public attitudes or behavior on scientific topics. The authors integrate these findings with more recent work on assessing the communication training landscape in North America. Most notably, the authors find that strategic communication is often not emphasized in these trainings, and skills are taught disconnected from these objectives. Directions for future research are discussed, especially within the context of expanding research–practitioner partnerships to advance science communication raining research.

In Chapter 2, Declan Fahy draws on the scholarship within the field of science studies, specifically the work of sociologists of science Harry Collins and Robert Evans, to discuss the role of expertise and experience of scientists within contemporary society. The chapter highlights the main tenants of Collins's and Evans's work in expertise-based communication, including the notion that scientists should communicate only from their specialist expertise, convey the nature of their expertise, explain how they adhere to the scientific ethos, and constantly evaluate how far their expertise extends. Fahy provides an overview of the key takeaways given the contemporary science and society interface, and discusses ways in which researchers and practitioners of communication training may integrate these takeaways.

In Chapter 3, Fred Balvert focuses on the historical importance of the relationship between research and industry, and argues that the contemporary theory and practice of science communication has not adequately accounted for this relationship. The chapter explores these themes through a historical and cross-cultural perspective, and discusses the role of various stakeholders, including researchers, research institutions, journalists, and private companies in this context. Balvert concludes the chapter by providing several conceptual models for the strategic communication of industry-funded research, and the implications for scientists, trainers, and professional science communicators.

Part I wraps up with a chapter by Emily Howell and Dominique Brossard that discusses the increased role of online media for science information, and the turn by science communicators to rely on social media to engage with peers, stakeholders, and interested publics. The authors discuss the opportunities and challenges for scientists' engagement in this new media landscape, highlighting what is known about best practices in successful science communication on social media. The chapter ends with a discussion about how science communication research and training can help inform best practices on social media, and the need for increased collaborations between researchers and practitioners.

In Part II of this volume, "Science Communication Training Design and Assessment," the focus shifts to several chapters on the practice of communication training, including the design and the development of training programs, as well as methods for evaluation and assessment.

In Chapter 5, Toss Gascoigne and Jenni Metcalfe describe their careers spanning nearly three decades providing communication training programs to scientists around the world. Starting with workshops in Australia, Gascoigne and Metcalfe have expanding to over 20 countries around the world. The authors describe the foundations of their workshop and how it has evolved over time to adapt to training scientists to communicate in new media environments and different sociocultural contexts. The chapter provides insight into the format of the skills that scientists are expected to master, as well as the challenges involved with integrating feedback and assessment into their own communication training program.

In Chapter 6, Tzipora Rakedzon describes an overview of training and assessment tools for scientific writing. Rakedzon describes the difficulties for scientists in shifting from writing for academics to lay audiences, and focuses on a number of best practices from existing training programs and educational contexts, as well as different assessment tools, including automated programs and rating rubrics. The chapter concludes with recommendations for assessment tools that can integrate into training programs, as well as directions for future research assessing the effectiveness of scientific writing.

In Chapter 7, Louise Kuchel integrates research in science education with science communication by providing an overview of best practices in educational approaches and how science communication trainers and researchers may benefit from this field. Kuchel describes how basic concepts in formal educational design including curriculum design, assessment, and learning is missing from science communication training, and provides recommendations for focusing attention on these concepts to support more effective communication training.

Part II concludes with a chapter by Yael Barel-Ben David and Ayelet Baram-Tsabari on the importance of evaluation within science communication training and the various tools developed to evaluate such training. In so doing, the authors describe the importance of defining learning goals and outcomes, and present a framework for evaluation based on the Human Resource Development (HRD) literature. The authors conclude the chapter with some of the inherent challenges in evaluation, and recommend several opportunities for researcher–practitioner collaborations.

Finally, Part III of this volume, "Future Directions for Science Communication Training," focuses on several novel frameworks for sustaining the communication training community as well as the engagement and outreach efforts of scientists.

In Chapter 9, Brenda MacArthur, Nicole Leavey, and Amanda Ng focus on the similarities between the growth of the science communication training field and the growth of the health communication training field. The authors discuss how the US healthcare system evolved to emphasize the value of communication training in formal medical training and the shift towards patient-centered approaches. The chapter concludes by providing an overview of best practices as well as pitfalls from the health communication training field, and exploring key lessons that the science communication training community can take away.

In Chapter 10, Brooke Smith describes the need to focus on a systems-based approach to cultivate and sustain the communication and engagement efforts of scientists. Using the infrastructure of the Metro system in Washington D.C. as an analogy, Smith emphasizes a number of key priorities that the scientific community should focus on to build an infrastructure that supports the role of training in helping scientists reach their communication and engagement goals.

In sum, these three sections provide a comprehensive overview of the research that has emerged within the field of science communication training. These chapters note the vibrant nature of the communication training field and the growth that it continues to experience around the world. Most importantly, these chapters link key findings on the role of the scientist, training program, and the broader scientific community in bridging the science–society interface, and the best practices that should be followed.

What is evident for the communication training community is the need to reflect on the growth of the field. To what end did these programs develop and how should the community orient itself toward sustaining a strategic focus moving forward? The extent to which this end is met, however, depends on understanding the effect that training has on scientists' communication efforts. The hope is that this volume will serve two important functions: First, the aim of this volume is to serve as a resource for science communication trainers about key issues within the field, and frameworks to consider either in starting or refining their own science communication training program. Second, the volume provides science communication researchers as well as practitioners with a number of different ways to approach fostering researcher–practitioner collaborations by highlighting directions for future research. Hopefully, this volume will contribute to the continued growth of communication training as a sub-discipline within the "science of science communication."

References

Cicerone R. 2010. Ensuring integrity in science. *Science, 327*(5966), 624.

Hall-Jamieson, K., Kahan, D., & Scheufele, D. A. (Eds.). 2017. *The Oxford Handbook of the Science of Science Communication.* New York: Oxford University Press.

Leshner A. I. 2015. Bridging the opinion gap. *Science, 347*(6221), 459.

Leshner A. I. 2003. Public engagement with Science. *Science, 299*(5609), 977.

National Science Board. 2018. *Science and technology: Public attitudes and public understanding* (Chapter 7). Science and Engineering Indicators.

Patrick, P. (Ed.). 2017. *Preparing Informal Science Educators: Lessons for Science Communication and Education.* New York: Springer.

Van der Sanden, M. C. A., & De Vries, M. J. (Eds.). 2016. *Science and Technology Education and Communication.* Rotterdam: Sense Publishers.

In Chapter 10, [illegible]

[illegible]

[illegible]

References

[illegible]

Part I

The scientist as a strategic communicator

1 Scientists, trainers, and the strategic communication of science

Nichole Bennett, Anthony Dudo, Shupei Yuan and John Besley

Introduction

In 2015, climate scientist Dr. Michael Mann appeared on HBO's *Real Time with Bill Maher* (Real Time with Bill Maher, 2015). Mann fumbled through the short conversation. He meandered through unclear talking points, displayed low energy, and failed to counter Maher's foul mood. In this interview, Mann missed an opportunity to contribute something noteworthy to the public sphere about climate change.

Strategic communicators scrutinizing this interview would have a lot to critique. They would question whether it made strategic sense for someone like Dr. Mann to appear on Maher's show. They would wonder what goals he was trying to achieve. And they would wonder if this platform, an argumentative late-night talk show, matched his communication goals.

Modern scientist communicators need more than good intentions when they engage the public. They need skills and strategy. This requires scientists to develop more sophisticated communication sensibilities, and trainers play a key role in this. Our research program investigates the role of strategy for science communicators and the trainers who support their efforts.

This chapter includes key insights and implications of our research program. First, we outline how scientists approach the public communication of science. Do these efforts include evidence of strategic thinking? Next, we describe how North American-based science communication trainers approach supporting scientists' communication efforts. Do they focus on strategic communication in their curricula? We conclude by highlighting implications for practice and by suggesting future research directions.

Scientists as communicators

Calls for scientists to communicate

There often appears to be a gulf between science and society, while the importance of science to society has never been greater. National Science Board (2014) data suggests that Americans have remained relatively positive about science in recent decades while also becoming increasingly concerned about specific technological issues such as genetic engineering and nuclear energy. Further, despite these positive views, they have continued to demonstrate limited knowledge and interest in scientific issues (Boudet et al., 2014; Gifford, 2011; Lee, Scheufele, & Lewenstein, 2005; NSB, 2010; Pew Research Center, 2008, 2015; van der Linden, Maibach, & Leiserowitz, 2015; Weber & Stern, 2011). Many of the scientific advances aimed at solving society's challenges (e.g., alternative energy, nanotechnology, synthetic biology, epigenetics, gene editing) raise sticky ethical, legal, and social questions, complicating and intensifying the public's responses to them (Dean, 2009; Leshner, 2003; Meredith, 2010; Priest, 2008). As scientific issues continue to impact the public, we likely need more quality communication between scientific experts and non-experts about these issues.

Scientists have unique access to knowledge, are seen as competent, and Americans want them to play a role in managing scientific issues (Fiske, Cuddy, & Glick, 2007; Funk, 2017; NSB, 2012). It therefore makes sense that they also occupy a central role in science communication. We need meaningful and proactive engagement by scientists with the public, especially in conversations involving risks to health and risks to environment (Bailey, 2010; Biegelbauer & Hansen, 2011; Cicerone, 2006; Corner, Markowitz, & Pidgeon, 2014; EU, 2002; Holt, 2015; Jia & Liu, 2014; Leshner, 2007; Lorenzoni et al., 2007; NASEM, 2016a; NRC, 1989; Pidgeon & Fischhoff, 2011; The Royal Society, 1985). Scientific leaders from major scientific organizations worldwide urge their colleagues to improve their communication skills and engage with the public (Cicerone, 2006, 2010; EU, 2002; Holt, 2015; Jia & Liu, 2014; Leshner, 2003, 2007, 2015; Reddy, 2009; Rowland, 1993; Department of Science and Technology: South Africa, 2014; The Royal Society, 1985). Some assert that scientists no longer enjoy the luxury of deciding *whether* to communicate with the public but instead must decide *how* they want to communicate (Donner, 2014; Pielke, 2007). This interest in the public communication of science generates critical questions about how scientists interface with science communication and with the public: (1) How often do scientists engage with the public? (2) What are the factors leading scientists to engage with the public? (3) How do scientists think about their science communication efforts?

How often do scientists communicate?

Despite ongoing calls for more engagement, many scientists already frequently engage in public communication activities, both directly with the public and indirectly through the media (Bauer & Jensen, 2011; Besley & Nisbet, 2013; Dudo et al., 2018; Hamlyn, Shanahan, Lewsi, O'Donoghue, & Burchell, 2015; Kreimer, Levin, & Jensen, 2011; Rainie, Funk, & Anderson, 2015; Torres-Albero, Fernandez-Esquinas, Rey-Rocha, & Martín-Sempere, 2011; The Royal Society, 2006). Scientists also value and plan to continue these outreach efforts (Dudo, Kahlor, Abi Ghannam, Lazard, & Liang, 2014; Dudo et al., 2018;

Martín-Sempere, Garzón-Garcia, & Rey-Rocha, 2008; Peters et al., 2008a; TRS, 2006). Many scientists report they got into science to make the world a better place (Pew Research Center, 2009) and view sharing what they know with others as a professional responsibility (Gascoigne & Metcalfe, 1997) or a way to raise public interest in science (DiBella, Ferri, & Padderud, 1991; Martín-Sempere et al., 2008; Peters et al., 2008a; TRS, 2006). Some scientists view getting coverage for their research as being important to career advancement (Kiernan, 2003; Milkman & Berger, 2014; Phillips, Kanter, Bednarczyk, & Tastad, 1991; Rainie et al., 2015; Shema, Bar-Ilan, & Thelwall, 2014). Others identify science communication as an enjoyable activity (Corrado, Pooni, & Hartfee, 2000; Dunwoody, Brossard, & Dudo, 2009; Martín-Sempere et al., 2008).

New communication technologies make the public communication of science more accessible and enticing to some scientists. The online media environment transforms how scientists interact with the public (Brossard & Scheufele, 2013; Linett, Kaiser, Durant, Levenson, & Wiehe, 2014). These platforms provide individual scientists with the opportunity to share their viewpoints on scientific issues for a more democratized science, engage in dialogue with a variety of stakeholders (Delborne, Schneider, Bal, Cozzens, & Worthington, 2103; Peters, 2013), or even contribute to the science itself (Einsiedel, 2014; Owens, 2014). This has led to enthusiastic pleas to scientists to embrace the use of social media to both become champions for the voice of science and reap the benefits of a social media presence for their own research, including increasing citation rates, improving broader impacts, and enhancing professional networking (Bik & Goldstein, 2013; Darling, 2014; Liang et al., 2014; Saunders et al., 2017; Van Eperen & Marincola, 2011; Wilkinson & Weitkamp, 2013).

What predicts scientists' willingness to engage?

To get more scientists involved in communication, we need to better understand the circumstances leading scientists to engage in public outreach. What processes underlie scientists' willingness to engage in high-quality public outreach activities? Much of the past research seeking to understand why scientists communicate lacked theory and relied on anecdotal evidences. Work within this theoretical framework demonstrates the empirical value of using the Theory of Planned Behavior (or the related Integrated Behavioral Model) as the theoretical framework for identifying the factors most associated with scientists' willingness to partake in the public communication of science. This theory describes how attitude, norms, and efficacy shape an individual's behavioral intentions and behaviors (Ajzen, 2017; Yzer, 2012). Research suggests the frequent importance of attitudes, efficacy, and norms in predicting scientists' willingness to engage (Besley, Oh, & Nisbet, 2013; Besley, Dudo, Yuan, & Lawrence, 2018; Poliakoff & Webb, 2007).

Social psychological variables

As might be expected, scientists with more positive attitudes toward engagement are more likely to conduct more engagement either when measured as a general affect toward engagement (Martín-Sempere et al., 2008; Poliakoff & Webb, 2007) or when measured as enjoyment (Besley et al., 2018; Dudo, 2013; Dunwoody et al., 2009). Positive attitudes toward the public may also increase scientists' willingness to engage; scientists who hold negative views about their expected audience are less likely to engage (Besley, 2014), although some studies found no significant relationship (Besley, Oh, & Nisbet, 2013; Besley, Dudo, & Storcksdieck, 2015). Attitude toward engagement was a more reliable indicator than attitude toward the audience in predicting engagement, suggesting past positive experiences lead to future engagement (Besley et al., 2018). To increase scientists' engagement activity, it will be worthwhile to emphasize how the activity is likely to be a positive experience or emphasize the benefits of engagement.

Social influence variables (i.e., norms) might also be expected to shape scientists' engagement efforts. We divide this variable into descriptive norms and subjective norms (Lapinski & Rimal, 2005). Descriptive norms are what scientists think their colleagues are doing (i.e., do scientists think their peers are engaging in public communication?), and subjective norms – or injunctive norms – are what scientists think their colleagues expect or would support (i.e., do scientists think their peers approve public communication?). Initial studies found scientists who believe their colleagues engage are more likely to say they intend to engage (Poliakaff & Webb, 2007), whereas more recent studies with larger sample sizes have found no such relationship (Besley, 2014; Besley et al., 2018). The "Sagan Effect," named after science popularizer Carl Sagan, reflects subjective norms. Sagan suffered professional setbacks because fellow scientists thought he could not be a serious scientist while also doing science outreach (Ecklund, James, & Lincoln, 2012; Fahy, 2015; Gascoigne & Metcalfe, 1997), but despite the cultural popularity of the term "Sagan Effect," research has not consistently found those associations (Besley & Nisbet, 2013; Besley & McComas, 2014; Besley et al., 2018; Dudo et al., 2014; Poliakoff & Webb, 2007). Normative beliefs seem to be less influential than thought.

We divide efficacy variables into three beliefs: self-efficacy (scientists' sense of their own skill at engaging), response efficacy (scientists' belief engagement can have a beneficial effect), and time (scientists' belief that they have the time to engage, which might be understood as an element of behavioral control). Scientists who believe they can do a reasonable job at public engagement (self-efficacy) are more willing to communicate (Besley et al., 2013; Besley, 2014; Dudo et al., 2014; Dunwoody et al., 2009; Poliakoff & Webb, 2007), as are scientists who believe engagement efforts can have desired impacts (response efficacy) (Besley et al., 2013, 2018; Besley, 2014). Past studies have found perceived time pressure does not predict engagement (Poliakoff & Webb, 2007), but more recent studies found time availability matters (Besley, 2014; Besley et al., 2018; Dudo, 2013). These findings for efficacy suggest concrete steps to encourage scientists to engage by improving their skills, their beliefs about impacts, and the time they have available to communicate.

Proponents of engagement could use these variables (attitudes, norms, efficacy) to identify and recruit scientists more likely to engage. For example, they could identify scientists more likely to achieve certain communication goals (e.g., targeting geographic locations or specific expertise) or ensure we support and encourage individuals from under-represented groups. It may be strategic to figure out how to get certain individuals to engage less, discouraging scientists who are aggressive (rude or condescending) who reinforce harmful stereotypes about scientists. These drivers may also be useful in efforts to change how scientists view communication. For example, if scientists who believe they are more skilled are also more willing to engage, then we can address this variable with science communication training.

While these studies provide insight into whether an individual scientist is likely to engage with someone outside of their area of research, they focus on quantity, not quality. We cannot expect every scientist to communicate, so it is important to emphasize the quality of the communication by those who do (Pearson, 2001). Examples of scientists communicating discussed above may represent one-way communication, which is less effective than meaningful, multi-party dialogue (Delli Carpini, Cook, & Jacobs, 2004; Grunig & Grunig, 2008). So, it is important to ask how scientists approach communication and investigate what objectives scientists include in their science communication activities.

Demographic variables

Demographic factors seem less important than social psychological factors in predicting a scientist's willingness to engage with the public. Some research suggests older scientists are somewhat more likely to engage with the public than younger scientists (Bentley & Kyvik, 2011; Besley et al., 2013, 2015; Crettaz von Roten, 2011; Dudo et al., 2018; Kreimer et al., 2011; Kyvik, 2005; Torres-Albero et al., 2011; The Royal Society, 2006) and male scientists are somewhat more likely to engage with the public than female scientists (Besley, 2014; Besley et al., 2013, 2018; Bentley & Kyvik, 2011; Crettaz von Roten, 2011; Kreimer et al., 2011; Torres-Albero et al., 2011). Other studies found the opposite pattern or no difference at all (Besley et al., 2018; Crettaz von Roten, 2011; Dudo, 2013; Ecklund et al., 2012; Jensen, 2011. Older scientists tend to hold leadership positions with outreach expectations (Rödder, 2012), or the security of tenure affords scientists the freedom and autonomy to engage (Dudo, 2013; Dudo et al., 2014). Yet other studies suggest initial early career increases in science communication activities with engagement tapering off with age (Besley & Oh, 2013; Besley et al., 2013, 2018).

Popular opinion predicts scientists in more applied fields will communicate more, but this claim lacks evidence. Some studies suggest certain fields engage with the public more than others, e.g., social scientists (Bauer & Jensen, 2011; Bentley & Kyvik, 2011; Jensen, 2011; Kreimer et al., 2011; Kyvik, 2005; Peters, 2013; Rainie et al., 2015), environmental scientists (Jensen, 2011; Rainie et al., 2015; Torres-Albero et al., 2011) or biologists/medical researchers (Besley et al. 2013; Marcinkowski, Kohring, Fürst, & Friedrichsmeier, 2013; Torres-Albero et al., 2011). But scientific field does not seem to be a primary driver of a scientist's engagement behavior (Besley et al., 2018; Ecklund et al., 2012).

Research suggests a cyclical relationship between past experiences, current views, and future behaviors (Ouellette & Wood, 1998), and a positive relationship exists between past engagement and willingness to take part in future engagement (e.g., Besley et al., 2018). In other words, if a scientist experiences communicating with the public, they generally wish to engage again.

How scientists approach communication

After a scientist has decided to communicate, she or he must decide how to communicate. Doing so effectively can benefit from making a series of strategic, thoughtful decisions. Ideally, scientists should have a clear long-term goal for their communication efforts in mind. This goal is likely some sort of behavior. This requires an understanding of who comprises a target audience for the goal and what makes that audience tick. Scientists should then think through what short-term communication objectives are likely to help them reach the long-term (behavioral) goal. Scientists must then choose the communication tactic – or tactics – most likely to help achieve the short-term objectives. These considerations represent basic tenets of strategic communication, and although they are common in fields like public relations (Hon, 1998) and health communications (Rice & Atkin, 2013), our research suggests they are not widely evident among scientist communicators.

The research team adapted this "Tactics-Objectives-Goals" model to describe the types of choices that communication researchers sometimes study and that scientist communicators could prioritize (Figure 1.1). In this model, "goals" are longer-term desired behavioral outcomes (Grunig & Hunt, 1984; Kendall, 1996) and "objectives" are shorter-term specific communication outcomes achieved along the way to the broader goal (Hon, 1998; Kendall, 1996). Objectives support goals, and "tactics" are what you do to achieve objectives. These big picture goals are easier for a scientist to define, as they are behaviors (or pseudo-behaviors) that originate from the scientists' research, resources, and personal interests. For example, do they want to help shape policy discussions? At what level (local, state, national)? Do they want to help influence the public's behavior, such as choosing STEM careers or making better medical or environmental decisions? One key is that it does not make sense to treat goals as the direct outcome of communication whereas communication objectives are the beliefs, feelings, and frames that result from communication activity (including communicative behavior). Identifying objectives can challenge scientists because these objectives may feel implicit or abstract, but for a science communicator,

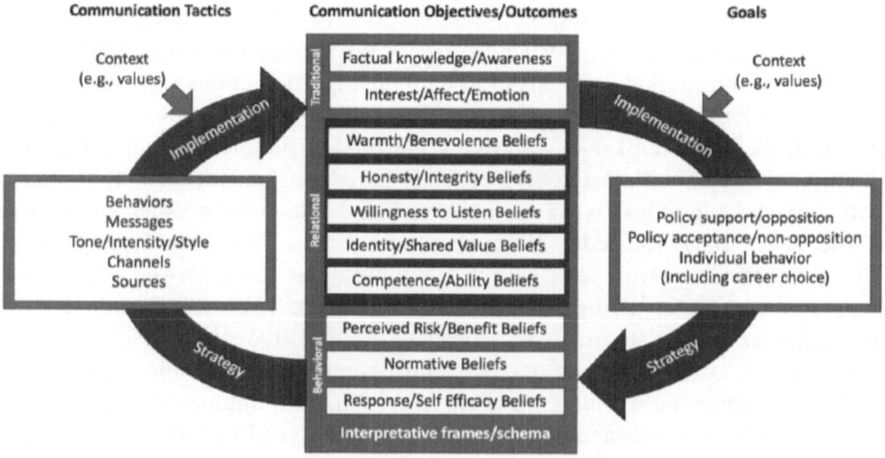

Figure 1.1 "Tactics-Objectives-Goals" Model.

prioritizing specific objectives means deciding where to put effort. Past research has identified a few categories of objectives that scientists may choose: increasing scientific knowledge, creating feelings such as excitement or curiosity, fostering various aspects of trust or relationships (e.g., warmth, competence, etc.), and framing a message to resonate with a specific audience. There are likely other objectives beyond these, but these objectives have been the most extensively researched.

Research on scientists' prioritization of objectives shows us that scientists continue to prioritize the objective of increasing knowledge (Bauer, Allum, & Miller, 2007; Besley & Nisbet, 2013; Besley et al., 2015; Besley, Dudo, Yuan, & Abi Ghannam, 2016; Burningham et al., 2007; Davies, 2008; Dudo & Besley, 2016; Fischhoff, 1995; Peters et al., 2008b; Petersen, Anderson, Allan, & Wilkinson, 2009; TRS, 2006) and often continue to operate on an incorrect assumption that increasing public knowledge about scientific facts and processes can substantially affect public support for science (the "deficit model") (Bauer et al., 2007; Besley & Nisbet, 2013; Dozier & Ehling, 1992; Fischhoff, 1995). Scientists place a lower priority on "non-informing" objectives, such as fostering excitement/interest, building trust, or framing issues to resonate) (Besley et al., 2015; Dudo & Besley, 2016). We argue that scientists would benefit from diversifying their science communication objectives and that trainers should help scientists develop the strategic skills needed to explicitly prioritize and achieve a broader range of communication objectives.

Empirical evidence continues to demonstrate that scientific knowledge has only a modest effect on attitudes toward scientific issues, and that work to increase public knowledge of science won't change behavior or increase support for policies (Allum, Sturgis, Tabourazi, & Brunton-Smith, 2008; Holmes, 1996; Hon, 1998). Besides having limited impact, this focus on educating the public may come at the expense of other (possibly more effective) communication objectives, like framing, building relational trust, or fostering efficacy (Dudo & Besley, 2016; Yuan, Besley, & Dudo, 2019). Positive beliefs about science and scientists are more likely to stem from high quality interactions with likable and engaging scientists who are willing to listen (Bauer et al., 2007; McComas, Arvai, & Besley, 2009), and improving the quality of science communication doesn't have to do with just getting better at explaining science facts, it has to do with fair and reciprocal ways to listen and understand each other's concerns (Bauer et al., 2007; Einsiedel, 2002; Hamlett & Cobb, 2006; Nep & O'Doherty, 2013). Research on trust tells us that it's helpful for those seeking support to have a reputation for caring about the needs of others (Fiske & Dupree, 2014), and attitudes related to trust are correlates of positive views about science (McComas & Besley, 2011). So, if a scientist is researching a topic out of a core desire to improve the world, it may help to prioritize communicating that to their audience through words and actions. Sparking interest and excitement can play a large role in spurring public motivations to seek future opportunities to learn about and engage with science (NRC, 2008), and interest-generating efforts may be core to shaping the public's views about science because they will drive additional attention, topic-related goal-setting, and learning (Hidi & Renninger, 2006), and can predict whether or not a person engages cognitively with materials or will ignore a message due to lack of motivation, from relevance or importance (Petty & Cacioppo, 1986). Allowing scientists to show their warmth, expertise, and willingness to listen to citizens may lead to more meaningful interactions between scientists and non-scientists, improving the reputation of science and scientists in the public's eye (Bies, 2005; Fiske & Dupree, 2014; Lauber, 1999; Webler, 2013). Framing issues to ensure that they resonate with a chosen audience also has a demonstrated effect on attitudes and behaviors, especially for science communication (Hart, 2011).

Why is there this disconnect between what objectives scientists prioritize and those that might be most effective for their goals? Educating the public is an objective that scientists feel most equipped to handle (Besley et al., 2015; Dudo & Besley, 2016; Edmondston, Dawson, & Schibeci, 2010), and they feel less skilled in demonstrating warmth or framing issues (Besley et al., 2015). Or there may be ethical concerns. Scientists may view dispelling pseudoscience and educating the public as more ethical and in line with the purity of science (Besley et al., 2015; Dudo & Besley, 2016) and may worry that trying to come across as warm and caring or framing an issue for their audience may come across as manipulative (Besley et al., 2015; Dudo & Besley, 2016; Holland, 2007).

Trainers may need to help scientists develop more of an appreciation for the importance of connecting with an audience and dispel some of the concerns about ethicality. It is likely impossible to communicate without shaping trust-related signals in a way that doesn't frame an issue in one or more ways, regardless of whether or not you are prioritizing this (Fiske & Dupree, 2014; Nisbet & Scheufele, 2009), and training may help scientists avoid accidentally coming across as cold or uncaring or framing issues in non-relevant ways. Trainers may play a role here in keeping scientists from misunderstanding framing (and other terms from the communication scholarship). And trainers can provide scientists with the skills and confidence to diversify their science communication objectives beyond educating the public.

When communicators make careful choices about objectives that have social science research backing up their effectiveness, better-quality communication can emerge. Scientists should be more likely to achieve their goals if they think more deeply about the choices they are making along the way and avoid ad hoc communication that isn't grounded in careful consideration of the short- and long-term impacts of their words and actions. The best predictors of how much scientists valued an objective was the degree to which they'd previously thought about it and the degree to which they see it as ethical (Besley et al., 2017a), and many worry that including strategy in science communication would introduce unethical advertising or public relations into the purity of science. Trainers may be able to play a role here in making compelling arguments for an objective's ethicality. No one should talk about motivations they don't possess, say they are listening when they are not, or frame issues in ways that aren't scientifically truthful.

While calls for more *quantity* of scientist-public interactions are not new (e.g., Bodmer, 1986; The Royal Society, 1985), leaders of the scientific community are increasingly calling for more *quality* science communication (e.g., Cicerone, 2010; Leshner, 2006, 2015; Napolitano, 2015; National Academy of Sciences, 2012, 2013; Pinholster, 2015). Research on scientists as science communicators gives us a clearer picture of how scientists engage with the public and the variables that drive these efforts, but research on scientists' communication objectives suggests they need support to translate this willingness into effective behavior. Trainers have emerged as a key conduit to address this need because they have the opportunity to connect communication scholarship and best practices to scientists willing to engage with the public and help them consider the strategic communication approach, asking not only how often and why scientists communicate but also aiming to understand what they hope to accomplish (Hon, 1998). Trainers can direct science communication efforts toward specific, intended outcomes (for the scientists, for their institutions, and for their audiences), and help scientists develop a more diverse set of communication objectives and related tactics (Besley & Tanner, 2011; Yuan et al., 2019). However, objectives may be hard to define for scientists because they aren't their overarching goal and aren't the immediate action. This is where communication scholarship comes in, and trainers can draw upon research on what objectives are appropriate for different types of goals and match them to tactics, helping scientists think more reflectively and strategically about their public communication efforts.

Training scientists to communicate

History and growth of professional training for scientists

For some scientists, talking about their research with non-experts seems to come naturally. But for many scientists, public communication skills are neither innate nor something they sought to cultivate. With the increased calls for scientists to improve their public communication (e.g., Leshner, 2015), how will scientists rise to the challenge? Enter professional science communication trainers, who are influencing the frequency and quality of scientist-public interactions (Besley & Tanner, 2011). Although groups that may have more regular contact with the public (e.g., journalists, regulators, health professionals) seem to receive more communication training than scientists (Besley & Tanner, 2011), training opportunities for scientists are growing (Besley et al., 2016; Brown, Propst, & Woolley, 2004; Fahy & Nisbet, 2011; Gold, 2001; Miller et al., 2009; Ossola, 2014; Peters et al., 2008a, 2008b; Reddy 2009; Russell 2006; Smith et al. 2013; Trench & Miller, 2012). One database created in 2018, for example, identified over 100 science communication training programs based in North America alone (http://strategicsciencecommunication.com).

The structure and curricula of these programs vary and include courses at universities, fellowships that include training, and stand-alone workshops with the goal of preparing scientists for interfacing with the public, working with media professionals, or using social media. These programs can run from a half-day up to a week or more, sometimes split over many sessions. Some high-profile examples in the United States include the Alan Alda Center for Communicating Science, the Center for Public Engagement with Science and Technology at the American Association for the Advancement of Science (AAAS), and COMPASS Science Communication.

Given the key role these training programs play in how scientists communicate with the public, it is useful to assess the nature of these programs and their impacts. This assessment involves examining descriptive aspects (e.g., comparing the number and expertise of program staff, comparing the number and expertise of their trainees, etc.), conceptual aspects (e.g., comparing topics emphasized in curricula, comparing learning objectives, etc.) and critical aspects (e.g., comparing how programs assess their contributions to trainees and the stakeholders with whom the trainees interact). These inquiries can reveal (in)consistencies across these programs' efforts, illuminate "best practices," and highlight opportunities for refinement.

Our research team regularly examines the North America-based science communication training ecosystem. We have been especially interested in examining the aims of their programming, particularly the extent to which it prioritizes key insights from strategic communication and science communication scholarship. We seek to address many research questions about the training programs, among them (1) What skills do they focus on teaching in their programs? (2) What types of communication goals and objectives do they highlight? (3) To what extent do their approaches incorporate best practices from the strategic communication research? And (4) What value do scientists see in these science communication training programs? To investigate these questions, we conducted qualitative interviews with science communication trainers across the U.S. (Besley et al., 2016; Yuan et al., 2017). We next discuss key findings from those interviews along with insights that emerged from similar projects conducted by a handful of other communication researchers.

Key findings from studies of science communication trainers

According to trainers, scientists seem to approach their engagement efforts with an array of communication goals. They come with personal goals (e.g., such as increasing visibility), or they come in with social goals (e.g., advocating for a specific policy) (Besley et al., 2016). Yet, our research found most training focuses on helping scientists communicate with one primary objective in mind: transmitting knowledge (Besley et al., 2016). In our interviews, only the objective of increasing knowledge was mentioned by most trainers without prompting. They rarely raised other communication objectives (e.g., building trust, conveying shared values) without prompting, and training curricula rarely incorporated these other objectives, even though trainers recognize the importance of these objectives (Yuan et al., 2017). As noted above, we find similar results in our surveys of scientists that show scientists' main objective in their science communication activities is to inform the public about science and defend science against misinformation (Dudo & Besley, 2016). Non-US studies also reflect this emphasis on communicating knowledge – programs focus on skills related to better explaining scientific phenomena (knowledge sharing) (Trench, 2012; Trench & Miller, 2012). Another of our studies found that whether a scientist had received training in the past and whether they felt they had the skills to communicate had little to do with what objectives they prioritize (Besley et al., 2017a). This is noteworthy because we hope that scientists with more communication skill and training develop a diversified set of communication objectives for their engagement activities. Another of our team's recent studies found that science communication scholars place more emphasis on designing engagement to facilitate scientist–public dialogue than scientists do (Yuan et al., 2019). This implies a potential disconnect between what communication scholars think trainers should teach scientists about communication with what professional trainers actually teach scientists about communication.

Science generates knowledge, and sharing knowledge represents a core piece of science communication. Yet relationship-building objectives also have value (e.g., fostering excitement about science, building trust in the scientific community, or reframing how people think about certain issues). But these "non-knowledge objectives" have a lower priority for scientists compared to sharing information about their research (Dudo & Besley, 2016), and few trainers explicitly name these non-knowledge objectives in interviews (Besley et al., 2016). Yet, trainers also said they believed many of the scientists they train want to raise public support for science (Besley et al., 2016). There's a clear problem with this logic of focusing on successful transmission of science information to increase public support: evidence suggests it fails. There is a minimal, albeit positive, association between science knowledge and science-related attitudes (Allum et al., 2008). And ample scholarship from the social sciences and the sub-field of science communication research details how fostering positive views about science requires more than correcting deficits in public knowledge (see Fischoff & Scheufele, 2014).

Our interviews also suggest that science communication trainers emphasize a subset of specific communication tactical skills in their curricula (Besley et al., 2016). Much of the advice scientists receive at these trainings tends to be about improving message clarity by removing jargon, developing presentation skills, storytelling and honing abilities associated with using communication technologies (e.g., audio, video, online publishing) (Baron, 2010; Besley et al., 2016; Hayes & Grossman, 2006; Olson, 2009; 2015). Further, much of the training is predominantly "journalistic" in character (i.e., focused on skills) and only minimally "strategic." That is to say, the training is less frequently focused on helping scientists determine their goals and objectives for engagement, identify target audiences, and select appropriate communication tactics. Non-US studies echo this pattern; scientists prefer practical communication skills training over training emphasizing scientific discourse (Miller, Fahy, & The ESConet, 2009). Training offers skills disconnected from the communication goals and objectives that trainers indicate they would be willing to prioritize (Besley et al. 2016; Yuan et al., 2017). Science communication scholars recommend that these training programs supplement their teaching of communication skills with information about basic communication theories and models (Besley & Tanner, 2011; Yuan et al., 2019).

When science communication training programs focus on building discrete communication skills disconnected from objectives, they risk improving skills in isolation from strategy. The associated concern is that science communication trainers focus too much on building technical skills and neglect to select specific communication objectives that theory and evidence show would help scientists achieve their long-term communication goals. So, although current training programs may be helping scientists become more technically proficient communicators, they may be communicating in ways that have little impact. This is, at best, inefficient and time-wasting, and at worst, damaging to the reputation of science itself. Skills may be the focus of these programs because abstract, theory-focused topics are more challenging to teach and are harder to connect with actually interacting with the public (Miller et al., 2009), whereas trainers perceive skills-based materials as easier to develop and more beneficial (Miller et al., 2009).

Treating science communication training like journalism training also raises ethical concerns. While journalists and scientists share many values, such as pursuit and commitment to the truth, it seems dishonest to think that scientist communicators are not choosing what to communicate by what impact they are hoping to have. Scientists usually have a direct interest in what they communicate about. Journalists are not supposed to promote a limited set of specific policy positions (Kovach & Rosenstiel, 2014) but scientists who engage with the public have science-focused goals (Besley et al., 2016). Since science communicators are not science journalists, training that emphasizes journalistic skills may be inappropriate or, at best, too narrow. Communication training curricula for scientists may be more appropriate and effective when it goes beyond conveying journalistic skills and also integrates well-established insights from other fields like public relations (Broom & Dozier, 1986; Dozier & Broom, 1995; Grunig & Grunig, 1989; Hon, 1998) and psychology (Locke & Latham, 1990, 2002, 2006).

So how do these findings translate to practice? There is an opportunity for science communication training programs to help scientists diversify or broaden the objectives they have for their communication efforts. Helping scientists realize that their engagement efforts can seek to achieve multiple objectives instead of one objective may be more enticing and beneficial in the same way offering a broader menu at a restaurant diversifies the types of meals one can choose from and enjoy (Besley et al., 2016). At the time of our research, science communication training seemed to place little emphasis on whether focusing on information-sharing helps scientists achieve the communication outcomes they seek. To address this disconnect, trainers could seek to customize and align training efforts to the specific communication goals a scientist is prioritizing. This may often mean that trainers need to help scientists understand the different types of objectives and goals available to them, and to help scientists identify which one/s are most appropriate for them to seek. Our interviews also suggest that trainers seldom help scientists identify the objectives and goals best suited to their communication style (Besley et al., 2016).

It is often easier for a scientist to communicate in ways that come naturally and hope for the best. And this knowledge-sharing form of communication is often the type of communication the broader scientific ecosystem seems to be rewarding. For example, science funders worldwide place an emphasis on improving researchers' abilities to communicate their results to non-academic audiences (Palmer & Schibeci, 2014), not to engage in a dialogue with members of the public or to show impact from engagement. Research suggests scientists who are more willing to engage, who think engagement makes a difference, and who view an objective as ethical are more willing to embrace strategic communication trainings (Besley et al., 2015). Scientists with these sensibilities may be more willing to approach engagement more strategically.

Thinking strategically means recognizing that our communication choices can have real consequences. Leaving time after a talk for real discussion, calling those with whom we disagree nasty names, or framing every disagreement as a war all impact the way the public perceives scientists (and science) (Hardy, Tallapragada, Besley, & Yuan, 2019). There is a danger that some scientists misconstrue being strategic as being dishonest, but we argue that effective strategic science communication relies on authenticity, like the science it represents. Ethical science communication trainers are unlikely to promote behaviors like pretending to be warm, fake listening, or inappropriate framing.

In sum, science communicators and science communication trainers are already providing outstanding training in key skills. They often focus on certain tactics that likely affect factors like trust even without drawing explicit connections to these objectives. Using accessible language and removing jargon from a talk may indeed help communicate that a scientist cares enough about their audience to put extra effort into making their research understandable to them. Storytelling becomes more than a way to convey information, it represents a social act. Effective public engagement is more likely to happen when there are high-quality interactions between people. The higher the quality of scientist–audience relationship, the more potential for mutually gratifying outcomes. Content matters, but it does not matter much unless a healthy dynamic for information exchange exists. Our interviews demonstrate that the science communication training community is already doing meaningful work. Integrating foundational strategic communication practices and insights from the social sciences will enrich their contribution to improving science communication.

Future directions

Research needed

Overall, we need more research on science communication training programs. There are lots of interesting exploratory stories that come out of our qualitative work on trainers that need to be followed up, whereas the work on the factors leading a scientist to engage is a bit more established. These studies could examine the type and scope of training programs available to science communicators, examining what they teach and why teach it, what educational resources they draw upon, and what training techniques they use. There is also a need to assess cross-sectional and longitudinal impacts of these training programs on individual scientists and the audiences with whom they communicate. Do trained scientists feel like they've improved? If so, how? Do the audiences with whom they engage perceive trained scientists differently than untrained scientists? Are the trained scientists warmer, more relatable, more trustworthy, etc.? If so, are these people more amenable/open to scientific ideas? We also need studies to investigate how scientists view training and what factors affect whether they participate in training. There may be organizational, socio-demographic, psychological, or other factors that make scientists more or less likely to see the value and participate in these science communication training programs. It may also be useful to study how scientists react to specific tactics expected to achieve their desired objectives. If they are willing to prioritize an objective, have they also thought about what it takes to achieve that objective?

Let us clarify again that our work has only looked at North American-based trainers, and we don't presume to generalize our findings across the globe. There is much more research needed here to investigate what are likely cultural differences across groups. We also need to know more about underrepresented groups, both as audience members and scientist communicators. Underrepresented groups' involvement in science communication is narrow and they experience exclusion due to feelings of cultural imperialism and powerlessness (Dawson, 2018). We need to reimagine the currently narrow and marginalizing "public" of science communication (Dawson, 2018). We need research and practical work to identify the effects of structural inequalities and pinpoint locations of exclusion/inclusion during the process of science communication (Dawson, 2018).

Implications for trainers

To get scientists to diversify beyond the one objective of filling knowledge gaps, we may want to get them to embrace specific science communication objectives in line with their larger goals. If we want science communication trainers to embrace strategic communication, trainers would benefit from support that helps them provide training in the framework of overall, long-term goals with intermediate objectives achievable along the way. The science communication scholarship could help provide trainers with the paths from specific objectives and long-term goals to actions, enabling scientists to communicate in ways that go beyond educating the public. From these paths, we could construct a typology connecting tactics, objectives, and goals for science communication activities. This typology would lay the foundation for a "curriculum menu" for trainers, who assist scientists in matching their own goal(s) with the skills and the communication activities necessary to reach objectives along the way to these goals. This way they could prioritize training on the skills to achieve these specific objectives and eventual goals.

Trainers may also be able to connect scientists with real-life opportunities to enact these objectives. One study showed scientists are willing to undertake training with the objective of engaging young audiences if there is a link to repeated access to youth through museums (Selvakumar & Storksdieck, 2013). This suggests that if trainers can connect scientists with tangible opportunities to achieve their objective, they may be more willing to see value in training that prepares them for those encounters.

This may be easier said than done, and we need to understand the challenges involved with a shift to strategic communication. There are likely reasons that scientists and trainers choose knowledge-building over other (possibly more effective) objectives, and we need a better understanding of scientists' and trainers' views of different objectives and goals. Trainers may be teaching some skills because they are easier to accomplish or they feel like they have quick wins for scientists (like removing jargon), leading trainers to choose tactics that feed ineffective objectives rather than choose more abstract and less (perceived) attainable objectives like building trust. Trainers may also be uncomfortable (ethics) or unfamiliar (skills) with teaching certain objectives as a route to scientists' goals. We need to face these challenges and concerns or help scientists find alternate routes to their goals.

We need more meaningful researcher-practitioner partnerships. In our vision, trainers and social scientists form a partnership to understand, monitor, and maximize training efforts. Science communication practitioners may be disconnected from relevant literature on their field (Miller, 2008), and the qualitative interviews with trainers suggest that they are not operating on best practices from the communication scholarship (Besley et al., 2016). We also need more concrete and practical tools from research for practitioners to use. For example, rather than hide scholarship away in journals, open-access multimedia tools about science communication best practices could share this information with science communication trainers. For this partnership to work the research community must also pay attention to the research needs of practitioners.

References

Ajzen, I. (2017). Icek Ajzen: Theory of planned behavior. Retrieved from http://people.umass.edu/aizen/tpb.html

Allum, N. C., Sturgis, P., Tabourazi, D., & Brunton-Smith, I. (2008). Science knowledge and attitudes across cultures: A meta-analysis. *Public Understanding of Science, 17*(1), 35–54.

Bailey, I. (2010). Creating a climate for change: Communicating climate change and facilitating social change – By Susanne C. Moser and Lisa Dilling. *Area, 42*(1), 133–134.

Bauer, M. W., Allum, N., & Miller, S. (2007). What can we learn from 25 years of PUS survey research? Liberating and expanding the agenda. *Public Understanding of Science, 16*(1), 79–95.

Bauer M. W. & Jensen, P. (2011). The mobilization of scientists for public engagement. *Public Understanding of Science, 20*(1), 3–11.

Baron, N. (2010). *Escape from the ivory tower: A guide to making your science matter.* Washington, DC: Island Press.

Bentley, P., & Kyvik, S. (2011). Academic staff and public communication: A survey of popular science publishing across 13 countries. *Public Understanding of Science, 20*, 48–63.

Besley, J. C. (2014). What do scientists think about the public and does it matter to their online engagement? *Science and Public Policy, 42*(2), 201–214.

Besley, J. C., Dudo, A., & Storksdieck, M. (2015). Scientists' views about communication training. *Journal of Research in Science Teaching, 52*(2), 199–220.

Besley, J. C., & Nisbet, M. (2013). How scientists view the public, the media and the political process. *Public Understanding of Science, 22*(6), 644–665.

Besley, J. C., & McComas, K. A. (2014). Fairness, public engagement and risk communication. In J. L. Arvai, & L., Rivers, (Eds.), *Effective Risk Communication* (pp. 108–123). New York, NY: Routledge/Earthscan.

Besley, J. C., & Tanner, A. H. (2011). What science communication scholars think about training scientists to communicate. *Science Communication, 33*(2), 239–263.

Besley, J. C., Oh, S. H., & Nisbet, M. (2013). Predicting scientists' participation in public life. *Public Understanding of Science, 22*(8), 971–987.

Besley, J. C., Dudo, A. D., Yuan, S., & Abi Ghannam, N. (2016). Qualitative interviews with science communication trainers about communication objectives and goals. *Science Communication, 38*(3), 356–381.

Besley, J. C., Dudo, A., Yuan, S., & Lawrence, F. (2018). Understanding Scientists' Willingness to Engage. *Science Communication, 40*(5), 559–590.

Besley, J. C., & Oh, S.-H. (2013). The combined impact of attention to the Deepwater Horizon oil spill and environmental worldview on views about nuclear energy. *Bulletin of Science, Technology & Society, 33*, 158–171.

Biegelbauer, P., & Hansen, J. (2011). Democratic theory and citizen participation: Democracy models in the evaluation of public participation in science and technology. *Science and Public Policy, 38*(8), 589–597.

Bies R. J. (2005). Are procedural justice and interactional justice conceptually distinct? In J. Greenberg & J. A. Colquitt, (Eds.), *Handbook of Organizational Justice* (pp. 59–85). Mahwah, NJ: Lawrence Erlbaum Associates.

Bik, Holly M. & Miriam C. Goldstein. 2013. An Introduction to Social Media for Scientists. *PLoS Biology, 11*, e1001535.

Bodmer, W. (1986). *The public understanding of science* (17th J. D. Bernal Lecture). Birkbeck College, London, England.

Boudet, H., Clarke, C., Bugden, D., Maibach, E., Roser-Renouf, C., & Leiserowitz, A. J. E. P. (2014). "Fracking" controversy and communication: Using national survey data to understand public perceptions of hydraulic fracturing. *Energy Policy, 65*, 57–67.

Broom, G. M., & Dozier, D. M. (1986). Advancement for public relations role models. *Public Relations Review, 12*(1), 37–56. doi:10.1016/S0363–8111(86)80039-X

Brossard, D., & Scheufele, D. A. (2013). Science, new media, and the public. *Science, 339*(6115), 40–41.

Brown, C. P., Propst, S. M., & Woolley, M. 2004. Report: Helping Researchers Make the Case for Science. *Science Communication, 25*, 294–303.

Cicerone, R. J. (2006). Celebrating and rethinking science communication. *The National Academy of Science: In Focus, 6*(1–2). Retrieved from www.infocusmagazine.org/6.3/president.html

Cicerone R. (2010). Ensuring integrity in science. *Science, 327*(5966), 624. pmid:20133539

Corner, A., Markowitz, E., & Pidgeon, N. (2014). Public engagement with climate change: The role of human values. *Wiley Interdisciplinary Reviews: Climate Change, 5*(3), 411–422.

Corrado, M., Pooni, K., & Hartfee, Y. (2000). The role of scientists in public debate. Retrieved from www.wellcome.ac.uk/About-us/Publications/Reports/Publicengagement/wtd003429.htm

Crettaz von Roten, F. (2011). Gender differences in scientist's public outreach and engagement activities. *Science Communication, 33*, 52–75.

Davies, S. R. (2008). Constructing communication: Talking to scientists about talking to the public. *Science Communication, 29*(4), 413–434.

Darling, Emily. (2014). It's Time for Scientists to Tweet. *The Conversation*. Retrieved from: https://theconversation.com/its-time-for-scientists-to-tweet-14658.

Dawson, E. (2018). Reimagining publics and (non) participation: Exploring exclusion from science communication through the experiences of low-income, minority ethnic groups. *Public Understanding of Science, 27*(7), 772–786.

Dean, C. (2009). *Am I making myself clear? A scientist's guide to talking to the public*. Cambridge, MA: Harvard University Press.

Department of Science and Technology: South Africa. (2014). Science engagement framework. Retrieved from www.saastec.co.za/science%20engagement%20framework%20final%20approved%20version.pdf

Delborne, J., Schneider, J., Bal, R., Cozzens, S., & Worthington, R. 2013.Policy Pathways, PolicyNetworks, and Citizen Deliberation: Disseminating the Results of World Wide Views on Global Warming in the USA. *Science and Public Policy*, *40*, 378–392.

Delli Carpini, M. X., Cook, F. L., & Jacobs, L. R. (2004). Public deliberation, discursive participation, and citizen engagement: A review of the empirical literature. *Annual Review of Political Science*, *7*, 315–344.

DiBella, Suzan M., Ferri, Anthony J., & Padderud, Allan B. (1991). 'Scientists' Reasons for Consenting to Mass Media Interview: A National Survey.' *Journalism Quarterly*, *68*, 740–749.

Donner, S. D. (2014). Finding your place on the science-advocacy continuum: An editorial essay. *Climatic Change*, *124*(1), 1–8.

Dozier, D. M., & Broom, G. M. (1995). Evolution of the manager role in public relations practice. *Journal of Public Relations Research*, *7*(1), 3–26. doi:10.1207/s1532754xjprr0701_02

Dozier D. M., & Ehling W. P. (1992). Evaluation of public relations programs: What the literature tells us about their effects. In J. Grunig (Ed.), *Excellence in public relations and communications management* (pp. 159–184). Hillsdale, NJ: Lawrence Erlbaum.

Dudo, A. (2013). Toward a model of scientists' public communication activity: The case of biomedical researchers. *Science Communication*, *35*, 476–501.

Dudo, A., & Besley, J. C. (2016). Scientists' prioritization of communication objectives for public engagement. *PloS One*, *11*(2).

Dudo, A., Besley, J., Kahlor, L. A., Koh, H., Copple, J., & Yuan, S. (2018). Microbiologists' Public Engagement Views and Behaviors. *Journal of Microbiology & Biology Education*, *19*(1).

Dudo, A., Kahlor, L., Abi Ghannam, N., Lazard, A., & Liang, M.-C. (2014). An analysis of nanoscientists as public communicators. *Nature Nanotechnology*, *9*, 841–844.

Dunwoody, S., Brossard, D., & Dudo, A. (2009). Socialization or rewards? Predicting U.S. scientist-media interactions. *Journalism & Mass Communication Quarterly*, *86*, 299–314.

Ecklund, E. H., James, S. A., & Lincoln, A. E. (2012). How academic biologists and physicists view science outreach. *PLoS One*, *7*, e36240.

Edmondston, J., Dawson, V., & Schibeci, R. (2010). are students prepared to communication? A case study of an Australian degree course in biotechnology. *International Journal of Science and Mathematics Education*, *8*(6), 1091–1108.

Einsiedel, E. F. (2002). Assessing a controversial medical technology: Canadian public consultations on xenotransplantation. *Public Understanding of Science*, *11*(4), 315–331.

Einsiedel, E. F. (2014). Publics and Their Participation in Science and Technology: Changing Roles, Blurring Boundaries. In B. Trench & M. Bucchi, (Eds.), *Routledge Handbook of Public Communication of Science and Technology*. Abingdon, Oxon: Routledge

European Union. (2002). Science and society: Action plan. Luxembourg: Office for Official Publications of the European Communities.

Fahy, D. (2015). *The new celebrity scientists: Out of the lab and into the limelight*. Lanham, MD: Rowman & Littlefield.

Fahy, D., & Nisbet, M. C. (2011). The science journalist online: Shifting roles and emerging practices. *Journalism*, *12*(7), 778–793.

Fischhoff, B. (1995). Risk perception and communication unplugged: Twenty years of process. *Risk Analysis*, *2*, 137–144.

Fischhoff, B., & Scheufele, D. A. (2014). The science of science communication II. *Proceedings of the National Academy of Sciences of the United States of America, 111*(Supplement 4), 13583–13584.

Fiske, S. T., Cuddy, A. J. C., & Glick, P. (2007). Universal dimensions of social cognition: Warmth and competence. *Trends in Cognitive Sciences, 11*, 77–83.

Fiske, S. T., & Dupree, C. (2014). Gaining trust as well as respect in communicating to motivated audiences about science topics. *Proceedings of the National Academy of Sciences of the United States of America, 111*(Supplement 4), 13593–13597.

Funk, C (2017) Mixed messages about public trust in science. Pew Research Center. Retrieved 3 March 2018 from www.pewinternet.org/2017/2012/2008/mixed-messages-about-public-trust-in-science/

Gascoigne, T., & Metcalfe, J. (1997). Incentives and impediments to scientists communicating through the media. *Science Communication, 18*, 265–282.

Gifford, R. (2011). The dragons of inaction: Psychological barriers that limit climate change mitigation and adaptation. *American Psychologist, 66*(4), 290–302.

Gold, B. D. (2001). The Aldo Leopold Leadership Program – Training Environmental Scientists to be Civic Scientists. *Science Communication, 23*, 41–49.

Grunig J. E., & Grunig L. A. (1989). Toward a theory of public relations behavior in organizations: Review of a program of research. In J. E. Grunig & L. A. Grunig (Eds.), *Public relations research annual* (Vol. I, pp. 27–66). Hillsdale, NJ: Lawrence Erlbaum Associates.

Grunig, J. E., & Grunig, L. A. (2008). Excellence theory in public relations: Past, present, and future. In A. Zerfass, B. Ruler & K. Sriramesh (Eds.), *Public relations research* (pp. 327–347). Wiesbaden, Germany: VS Verlag für Sozialwissenschaften.

Grunig, J., & Hunt, T. (1984). *Managing public relations*. New York, NY: Holt, Rinehart & Winston.

Hamlett, P. W., & Cobb, M. D. (2006). Potential solutions to public deliberation problems: Structured deliberations and polarization cascades. *Policy Studies Journal, 34*(4), 629–648.

Hamlyn, B., Shanahan, M., Lewsi, H., O'Donoghue, T., & Burchell, K. (2015). Factors affecting public engagement by researchers: A study on behalf of a consortium of UK public research funders. Retrieved from www.wellcome.ac.uk/stellent/groups/corporatesite/@msh_grants/documents/web_document/wtp060033.pdf

Hart, P. S. (2011). One or many? The influence of episodic and thematic climate change frames on policy preferences and individual behavior change. *Science Communication, 33*(1), 28–51.

Hayes, R., & Grossman, D. (2006) *A Scientist's Guide to Talking with the Media: Practical Advice from the Union of Concerned Scientists*. New Brunswick, NJ: Rutgers University Press.

Hidi, S., & Renninger, K. A. (2006). The four-phase model of interest development. *Educational Psychologist, 41*(2), 111–127.

Holland, E. (2007). The risks and advantages of framing science. *Science, 317*(5842), 1168–1170.

Holmes, P. A. (1996, April). Evaluation: What is more important than who. *Inside PR*, 2.

Hon, L. C. (1998) Demonstrating effectiveness in public relations: Goals, objectives, and evaluation. *Journal of Public Relations Research, 10*, 103–135.

Holt, R. D. (2015). Why science? Why AAAS? *Science, 347*, 807–807.

Jensen, P. (2011). A statistical picture of popularization activities and their evolutions in France. *Public Understanding of Science, 20*, 26–36.

Jia, H., & Liu, L. (2014). Unbalanced progress: The hard road from science popularisation to public engagement with science in China. *Public Understanding of Science, 23*, 32–37.

Kendall, R. L. (1992) *Public Relations Campaign Strategies: Planning for Implementation.* New York, NY: HarperCollins.

Kiernan, V. (2003). Diffusion of news about research. *Science Communication, 25*(1), 3–13.

Kovach B., Rosenstiel T. (2014). *The elements of journalism: What newspeople should know and the public should expect* (Rev. updated 3rd ed.). New York, NY: Three Rivers Press.

Kreimer, P., Levin, L., & Jensen, P. (2011). Popularization by Argentine researchers: The activities and motivations of CONICET scientists. *Public Understanding of Science, 20*, 37–47.

Kyvik, S. (2005). Popular science publishing and contributions to public discourse among university faculty. *Science Communication, 26*, 288–311.

Lauber, T. B. (1999). Measuring fairness in citizen participation: a case study of moose management. *Society & Natural Resources, 12*(1): 19–37.

Lapinski, M. K., & Rimal, R. N. (2005). An explication of social norms. *Communication Theory, 15*, 127–147.

Lee, C.-J., Scheufele, D. A., & Lewenstein, B. V. 2005. Public Attitudes Toward Emerging Technologies: Examining the Interactive Effects of Cognitions and Effect on, Public Attitudes Toward Nanotechnology. *Science Communication, 27*, 240–267.

Leshner, A. I. (2006). Science and public engagement. *The Chronicle of Higher Education.* Retrieved from http://chronicle.com/article/SciencePublic-Engagement/25084

Leshner, A. I. (2007). Outreach training needed. *Science, 315*, 161.

Leshner, A. I. (2015). Bridging the opinion gap. *Science, 347*(6221), 459. doi:10.1126/science.aaa7477

Leshner, A. I. (2003). Public engagement with Science. *Science, 299*(5609), 977. Pmid:12586907

Liang, X., Su, L. Y. F., Yeo, S. K., Scheufele, D. A., Brossard, D., Xenos, M., … & Corley, E. A. (2014). Building Buzz: (Scientists) Communicating Science in New Media Environments. *Journalism & Mass Communication Quarterly, 91*(4), 772–791.

Linett, P., Kaiser, D., Durant, J., Levenson, T., & Wiehe, B. (2014). The evolving culture of science engagement. An exploratory of the Massachusetts institute of technology and culture kettle. Report of Findings: September 2013 workshop.

Locke, E. A., & Latham, G. P. (1990). *A theory of goal setting and task performance.* Englewood Cliffs, NJ: Prentice Hall.

Locke, E. A., & Latham, G. P. (2002). Building a practically useful theory of goal setting and task motivation. *American Psychologist, 57*, 705–717. doi:10.1037//0003-066X.57.9.705

Locke, E. A., & Latham, G. P. (2006). New directions in goal-setting theory. *Current Directions in Psychological Science, 15*(5), 265–268. doi:10.1111/j.1467-8721.2006.00449.x

Marcinkowski F., Kohring M., Fürst S., & Friedrichsmeier A. (2013). Organizational influence on scientists' efforts to go public: An empirical investigation. *Science Communication, 36*, 56–80.

Martín-Sempere, M. J., Garzón-Garcia, B., & Rey-Rocha, J. (2008). Scientists' motivation to communicate science and technology to the public: Surveying participants at the Madrid Science Fair. *Public Understanding of Science, 17*, 349–367.

McComas. K. A., & Besley J. C. (2011). Fairness and nanotechnology concern. *Risk Analysis, 31*(11): 1749–1761.

Meredith, D. (2010). *Explaining Research: How to Reach Key Audiences to Advance Your Work.* New York, NY: Oxford University Press.

Miller, S. (2008). So where's the theory? On the relationship between science communication practice and research. In D. Cheng, M. Claessens, N. R. J. Gascoigne, J. Metcalfe, B. Schiele, & S. Shi (Eds.), *Communicating science in social contexts* (pp. 275–287). New York, NY: Springer Science + Business Media.

Miller, S., Fahy, D., & The ESConet Team. (2009). can science communication workshops train scientists for reflexive public engagement? The ESConet experience. *Science Communication, 31*(1), 116–126.

Milkman, K. L., & Berger, J. (2014). The science of sharing and the sharing of science. *Proceedings of the National Academy of Sciences, 111*(Suppl. 4), 13642–13649.

Napolitano J. (2015). Why more scientists are needed in the public square. The Conversation. Retrieved from https://theconversation.com/why-more-scientists-are-needed-in-the-public-square-46451

National Academy of Sciences. (2012). The science of science communication. Retrieved from www.nasonline.org/programs/sackler-colloquia/completed_colloquia/science-communication.html

National Academy of Sciences. (2013). The science of science communication II. Retrieved from www.nasonline.org/programs/sackler-colloquia/completed_colloquia/agenda-science-communication-II.html

National Research Council. (2008). *Public participation in environmental assessment and decision making.* Washington, DC: National Academy Press.

National Science Board. (2014, July). Chapter 7, Science and technology: Public attitudes and public understanding. Science and engineering indicators. Retrieved from http://www.nsf.gov/statistics/seind12/

Nep, S., & O'Doherty, K. (2013). Understanding public calls for labeling of genetically modified foods: Analysis of a public deliberation on genetically modified salmon. *Society & Natural Resources, 26*(5), 506–521.

Olson, R. (2009). *Don't be such a scientist: Talking substance in an age of style.* Washington, DC: Island Press.

Olson, R. (2015). *Houston, we have a narrative: Why science needs story.* Chicago: University of Chicago Press.

Ouellette, J. A., & Wood, W. (1998). Habit and intention in everyday life: The multiple processes by which past behavior predicts future behavior. *Psychological Bulletin, 124*(1), 54.

Ossola, A. 2014. How Scientists are Learning to Write. *The Atlantic*, December 12.

Owens, S. 2014.How Reddit Created the World's Largest Dialogue Between Scientists and the General Public. *Tech, Media, and Marketing.* Retrieved on December 9, 2014 from www.simonowens.net/how-reddit-created-the-worlds-largest-dialogue-between-scientists-and-the-general-public

Palmer, S. E., & Schibeci, R. A. (2014). What conceptions of science communication are espoused by science research funding bodies?. *Public Understanding of Science, 23*(5), 511–527.

Pearson, G. (2001). Participation of scientists in public understanding of science activities: The policy and practice of the U.K. Research Councils. *Public Understanding of Science, 10*, 121–137.

Peters, H. P. (2013). Gap between science and media revisited: Scientists as public communicators. *Proceedings of the National Academy of Sciences of the USA, 110*, 14102–14109.

Peters, H. P., Brossard, D., de Cheveigne, S., Dunwoody, S., Kallfass, M., Miller, S., & Tsuchida, S. (2008a). Science-media interface: It's time to reconsider. *Science Communication, 30*, 266–276.

Peters, H. P., Brossard, D., de Cheveigne, S., Dunwoody, S., Kallfass, M., Miller, S., & Tsuchida, S. (2008b). Science communication: Interactions with the mass media. *Science, 321*, 204–205.

Petersen, A., Anderson, A., Allan, S., & Wilkinson, C. (2009). Opening the black box: Scientists' views on the role of the news media in the nanotechnology debate. *Public Understanding of Science, 18*(5), 512–530.

Petty, R. E., & Cacioppo, J. T. (1986). The elaboration likelihood model of persuasion. Springer

Pew Research Center for the People & the Press. (2008, August). Audience segments in a changing news environment: Key news audiences new blend online and traditional sources. Retrieved from http://people-press.org/reports/pdf/444.pdf

Pew Research Center. (2009). Public praises science; scientists fault public, media. Retrieved January 3, 2015 from www.people-press.org/2009/07/09/public-praises-science-scientists-fault-public-media/.

Pew Research Center. (2015, January). Public and Scientists Views on Society.

Phillips, D. P., Kanter, E. J., Bednarczyk, B., & Tastad, P. L. (1991). Importance of the lay press in the transmission of medical knowledge to the scientific community. *New England Journal of Medicine, 325*(16), 1180–1183.

Pinholster G. (2015). AAAS unveils Alan I. Leshner Leadership Institute. American Association for the Advancement of Science Newsroom. Retrieved from www.aaas.org/news/aaas-unveils-alan-i-leshner-leadership-institute

Pielke, R. A. (2007). *The honest broker: Making sense of science in policy and politics*. New York: Cambridge University Press.

Poliakoff, E., & Webb, T. L. (2007). What factors predict scientists' intentions to participate in public engagement of science activities. *Science Communication, 29*(2), 242–263.

Priest, S. (2008). Biotechnology, nanotechnology, media, and public opinion. In K. David & P. B. Thompson (Eds.), *What can nanotechnology learn from biotechnology? Social and ethical lessons for nanoscience from the debate over agrifood biotechnology and GMOs* (pp. 221–234). Burlington, MA: Elsevier.

Rainie, L., Funk, C., Anderson, M. (2015). How scientists engage the public. Retrieved from www.pewinternet.org/2015/02/15/how-scientists-engage-public/

Real Time with Bill Maher. (2015, August). Real Time with Bill Maher: Dr. Michael Mann on Climate Change – August 7, 2015 (HBO) [Video File]. Retrieved from www.youtube.com/watch?v=nZ2cCPRS-Q8

Reddy C. (2009). Scientist citizens. *Science, 323*(5929): 1405.

Rice, R. E., & Atkin, C. K. (2013). *Public communication campaigns* (4th edition). Thousand Oaks, CA: SAGE.

Rödder, S. (2012). The ambivalence of visible scientists. In S. Rödder, M. Franzen & P. Weingart (Eds.), *The sciences' media connection: Public communication and its repercussions* (pp. 155–177). Dordrecht, Netherlands: Springer.

Rowland, Frank S. (1993). President's Lecture: The Need for Scientific Communication with the Public. *Science, 260*, 1571–1576.

The Royal Society. (1985). *The public understanding of science* ("The Bodmer Report"). London, UK: The Royal Society.

Russell, Christine. (2006). Covering Controversial Science: Improving Reporting on Science and Public Policy. Joan Shorenstein Center on the Press, Politics and Public Policy.

Saunders, M. E., Duffy, M. A., Heard, S. B., Kosmala, M., Leather, S. R., McGlynn, T. P., … & Parachnowitsch, A. L. (2017). Bringing ecology blogging into the scientific fold: measuring reach and impact of science community blogs. *Royal Society Open Science*, *4*(10), 170957.

Selvakumar, M., & Storksdieck, M. (2013). Portal to the public: Museum educators collaborating with scientists to engage museum visitors with current science. *Curator: The Museum Journal*, *56*(1), 69–78.

Shema, H., Bar-Ilan, J., & Thelwall, M. (2014). Do blog citations correlate with a higher number of future citations? Research blogs as a potential source for alternative metrics. *Journal of the Association for Information Science and Technology*, *65*(5), 1018–1027.

Smith, B., Baron, N., English, C., Galindo, H., Goldman, E., McLeod, K., …, & Neeley, E. 2013. 'COMPASS: Navigating the Rules of Scientific Engagement.' *PLoS Biology*, *11*, e1001552.

Torres-Albero, C., Fernandez-Esquinas, M., Rey-Rocha, J., & Martin-Sempere, M. J. (2011). Dissemination practices in the Spanish research system: Scientists trapped in a golden cage. *Public Understanding of Science*, *20*, 12–25.

Trench, B. (2012). Vital and vulnerable: Science communication as a university subject. In B. Schiele, M. Claessens & S. Shi (Eds.), *Science Communication in the World–practices, theories and trends* (pp. 241–257). Springer.

Trench, B., & Miller, S. (2012). Policies and practices in supporting scientists' public communication through training. *Science and Public Policy*, *39*(6), 722–731.

Van Eperen, L., & Marincola, F. M. (2011). How Scientists Use Social Media to Communicate Their Research. *Journal Transl Med*, *9*, 10.1186

van der Linden, S., Maibach, E., & Leiserowitz, A. (2015). Improving public engagement with climate change: Five "best practice" insights from psychological science. *Perspectives on Psychological Science*, *10*(6), 758–763.

Weber, E. U., & Stern, P. C. (2011). Public understanding of climate change in the United States. *American Psychologist*, *66*(4), 315–328.

Webler T. (2013). Why risk communicators should care about the fairness and competence of their public engagement processes. In: J. L. Arvai & L. Rivers (Eds.), *Effective risk communication* (pp. 124–141). London: Earthscan.

Wilkinson, C., & Weitkamp, E. (2013). A case study in serendipity: environmental researchers use of traditional and social media for dissemination. *PloS One*, *8*(12), e84339.

Yuan, S., Oshita, T., Abi Ghannam, N., Dudo, A., Besley, J. C., & Koh, H. E. (2017). Two-way communication between scientists and the public: a view from science communication trainers in North America. *International Journal of Science Education, Part B*, *7*(4), 341–355.

Yuan, S., Besley, J. C., & Dudo, A. (2019). A comparison between scientists' and communication scholars' views about scientists' public engagement activities. *Public Understanding of Science*, *28*(1), 101–118.

Yzer, M. C. (2012). The integrative model of behavioral prediction as a tool for designing health messages: Theory and practice. In H. Cho (Ed.), *Designing messages for health communication campaigns: Theory and practice* (pp. 21–40). Thousand Oaks, CA: Sage.

2 What studies of expertise and experience offer science communication training

Declan Fahy

When I was part of a European network in the 2000s that delivered communication training workshops to scientists, we left our most important and problematic module for last. The module was titled Science in Culture. We delivered it as a final session of the workshop, after the workshop trainees, mainly early-career scientists, had taken modules in basic communication skills, such as writing for popular audiences and preparing for media interviews. Science in Culture, in contrast, was a more conceptual and discursive module. It took as its premise the idea that science, as an integral part of modern life, cannot be understood as separate from its rich and dynamic cultural environment. As we noted in our description of the module, "although science has tried to protect itself from outside influences, science and science communication may be influenced by political, social and economic interests within various contexts. In this way, science can be seen as embedded in culture." We in the European Science Communication Network (ESConet) believed scientists with a sophisticated understanding of the place of science in culture would have a clearer view of the distinctive contribution they could make to public life.[1]

The response from scientists to the module was polarized. Our end-of-workshop qualitative feedback from trainees found that as many participants found Science in Culture intellectually stimulating as believed it was a waste of time that would have been better spent on learning practical skills, such as undergoing mock interviews or learning how to write for non-specialists (Miller, Fahy, and the ESConet team, 2009). Nevertheless, we continued to deliver the module, as it fitted with our network's framework for thinking about science communication – a framework that viewed communication as public dialogue that fostered civic engagement with science. The modules, moreover, marked a culmination of several years' work by researchers to create 12 original teaching modules that were delivered in our workshops (see ENSCOT, 2003). As part of the process of creating the modules, we drew on ideas from science communication and the field variously known as science studies, the sociology of scientific knowledge, or science and technology studies (STS). The field has done much since its origins in the 1970s to bring scholarly attention to the place of science in culture, and we argued that its scholarship provided a reservoir of useful knowledge that could be used to help scientists communicate in broader culture.

But since our work with ESConet, the field of science communication training has not engaged to the same degree, to my mind, with ideas from science studies. According to trainers, programs focus overwhelmingly on developing practical media skills (Besley, Dudo, Yuan, & Abi Ghannam, 2016). When scholarly knowledge is incorporated, it is drawn mostly from the public persuasion research tradition in communication studies, examining how communication objectives can be clarified, how audiences can be segmented, and how messages can be tailored to resonate with those audiences (Dudo & Besley, 2016). When scientists reflect on their own motivations for engaging with wider audiences, they prioritize communication that seeks to defend science from perceived misinformation or to educate the public about science, a public that scientists tend to view as irrational and misguided (Dudo & Besley, 2016). One result of these trends is that scientists can come to training programs seeking the skills to become, in effect, public defenders of science, and trainers seek to give the skills to engage in a form of strategic science communication. Yet missing from this focus on practical skills and strategic messaging is a deep understanding of the role of science in culture, an understanding that needs to take into account the cultural dynamics in which contemporary science operates, and the role scientists can play in such a dynamic culture.

In this chapter I will argue that one influential strand of contemporary science studies scholarship offers a way to think about the role of science in culture. I will argue that the collaborative work of sociologists of science Harry Collins and Robert Evans on the nature of scientific expertise and experience has foundational implications for science communication training. Synthesizing a strand of their work, I will argue that training researchers to communicate about science means training them to communicate about their expertise, experience and ethos as scientists. This way of viewing science communication training is vital because contemporary scientists, in the West, are communicating in a cultural environment where their expertise is challenged, where their expertise interacts with other forms of knowledge, and where citizens expect to participate democratically in debates over science and technology. A focus on expertise and experience will help put communication training, as a field, on a coherent conceptual foundation. A focus on expertise and experience will also help trainers make sense of the two hard problems that scientists who communicate must address. Does their scientific expertise and experience give them legitimacy to contribute to public discussion? And how far does their technical expertise extend when discussing science-related social problems?

The value of science studies for communication training

The neglect of science studies is not particular to training programs. As the science policy scholars Michael Crow and Daniel Sarewitz (2013, p. 4) argued, ideas from science studies have achieved only a marginal presence in the academy and government, even though they can make an important contribution to the policy world. The field of science studies has also been satirized or dismissed. In a recent example, the cognitive scientist and public intellectual Steven Pinker in his *Enlightenment Now* (2019, p. 396) put science studies in quotation marks, denoting the less-than-serious nature of some scholars in "science studies". For him, science studies – which has its different schools of thought and a spectrum of epistemological positions that range from unreconstructed positivism to extreme relativism – has been equated exclusively with postmodernism. In addition, the philosopher of science Helena Sheehan (2007, p. 207) argued that science studies had retreated from major social issues into narrow disciplinary debates, becoming too esoteric and insular, beset with what she called "mini-debates of micro-tendencies." In response to these criticisms, a task for science communication training scholars is to focus attention on useful research science studies offers. As an entry on science studies in an encyclopedia of science communication noted:

> By recognising the truth and the limitations of common assumptions about science and technology and getting beyond oversimplified conceptions regarding them, STS helps illuminate what is at stake in and what is required for the deliberate use of science and technology to achieve social goals.
>
> (Neeley, 2010, p. 742)

Collins and Evans in a recent strand of their scholarship have argued that expertise and experience are central to understanding the role of science in culture and society. In their words, they have moved "from evaluating science as a provider of truth to analyzing the meaning of the expertise upon which the practice of science and technology rests" (2007, p. 2). They argue that expertise is real, that expertise is hard earned, and that expertise is the foundation of scientists' public legitimacy. In this chapter, I build on the work of Collins and Evans to show how expertise and experience are central to scientists' public communication and why, by extension, should be central to scientists' communication training. In this chapter, I draw on three of their works. The first is their co-authored paper in *Social Studies of Science* (Collins & Evans, 2002) that set out their view that expertise and experience have become central to contemporary debates about science and society. The second is their joint-authored *Rethinking Expertise* (Collins & Evans, 2007) that expanded and deepened these ideas. The third is *Are We All Scientific Experts Now?* (2014), in which Collins distilled ideas from earlier works into a book accessible to a wide audience. (And, short answer: No, we are not all scientific experts now.)

Collins is a sociologist of science who has much to offer communication training. First and importantly, he knows science deeply. As a sociologist, he has been at the coalface of one field since its origins, having studied since the early 1970s the work of physicists who study gravitational waves, emitted when stars explode or collide. He detailed the history and sociology of the field in his book *Gravity's Shadow: The Search for Gravitational Waves* (2004), becoming so immersed in the field that he developed advanced-level factual knowledge as well as obtaining a degree of the specialists' tacit knowledge, the knack experts have of being able to do their specialist tasks without being able to explain how they do it. "I have spent decades studying the sociology of the detection of gravitational waves and hanging around with the gravitational-wave physics community," he wrote (2014, p. 69). "I certainly cannot 'do' gravitational-wave physics but I decided to expose myself to a test that would indicate if I had a grasp of the tacit knowledge belonging to the specialists." In the test he and a professional gravitational-wave physicist were asked identical technical questions. The anonymized results were then evaluated by nine gravitational-wave physicists, who had to identify the real specialist and Collins. In the evaluation, seven physicists could not tell who the real physicist was, and two identified Collins as the specialist. The result was written up in the scientific journal *Nature* (Giles, 2006) under the heading "Sociologist Fools Physics Judges."

Second, Collins's expertise has gone beyond advanced factual knowledge. As part of his immersion in the sub-field, Collins came to understand the social and cultural worlds of gravitational-wave physics. He was able to examine how scientific ideas in the field originated and developed and came to be accepted or rejected. In the process, he developed a feel for the inner workings of the field, the hard-to-define social and cultural factors that only someone immersed in a discipline could begin to understand. As a result, he argued in his work that social factors were crucial to scientific knowledge production – social factors that included how researchers evaluated each other's work and reputation, how they discussed their theories, interpreted their experimental results, and calibrated their instruments. But, Collins (2014) noted, even though social factors were involved, science created reliable knowledge of the natural world – knowledge that has served as the basis for personal and political action (for more on the concept of reliable knowledge, see Ziman, 1991).

Third, Collins, with Evans, has argued that science deserves its central position in Western culture. Their work aimed, in part, to defend science's special epistemological status, its unique claim to be a method of inquiry that has produced reliable knowledge of the natural world. They also argued that science was a central part of culture. It deserved its special place, in their view, because scientists have been committed to a set of values that can be considered the scientific ethos. These values, or norms, include those developed by sociologist of science Robert Merton, such as "universalism" in that claims are judged without reference to the personal attributes of whoever makes the claims, "organized skepticism" in that work undergoes critical peer scrutiny, and "disinterestedness" in that scientists act for the benefit of science as a whole, not their own personal interest, as well as the values of honesty and integrity (Collins, 2014, pp. 127–128; Merton, 1973). "The scientific ethos," wrote Collins (2014, p. 132), "may be the most valuable contribution of science to society".

The scientific ethos, for Collins (2014, p. 126), can be most clearly seen in the work of what he called "the heartlands of professional science." When he and Evans analyzed the scientific community, they were not examining visible public scientists who were fixtures in the media (Goodell, 1977), or celebrity scientists who came to symbolize science in popular culture (Fahy, 2015), or scientists who were active players in the policy process (Pielke, Jr., 2007). They were discussing, instead, the theorists who aimed to support or disprove their ideas, and the experimentalists who sought to determine the fabric of the natural world. These researchers constituted the vast ranks of professional scientists. They were exemplified, for Collins, by the gravitational-wave physics community, which in his experience have been committed, with very few exceptions, to the scientific ethos. In my experience of training also, it was exactly these scientists from the heartlands that came to our workshops, motivated by a desire to enhance the public understanding of science. They were not, in my experience, engaged in a sophisticated strategic communication campaign on behalf of science, but instead wanted to learn how to convey vividly the scope and limits of science as a way of understanding and explaining the natural world. It has been these scientists who have constituted the core constituency for communication training. And so training should focus heavily on what these scientists have to offer the public understanding of science, and what challenges they are likely to face as they communicate in public.

The third wave of science studies: studies of expertise and experience

The first point to understand, based on Collins and Evans's work, is that science, and therefore science communication, is situated in an ever-changing historical context. The public image of science at a given historical moment, Collins (2014) argued, can be understood using the term *zeitgeist*, a concept that refers to the prevailing environment of an era, the general cultural and intellectual climate of an age, the popular imagination existing at a particular time and place. That intellectual climate influences how people make sense of science. To trace changes in the popular imagination towards science, Collins and Evans (2002) traced what they argued were the large-scale changes in the field of science studies since the middle of the last century. Changes in this field provided a window into the changes in wider society, they argued, and the history of the field can be broadly categorized into three waves, with each wave reflecting the changing relationship between science and wider society.

The first wave of science studies started in the 1950s, in an era of new drugs, new materials, and new technological optimism for a post-war future. The first wave saw science as the pre-eminent form of knowledge. In the first wave, scholars outside science had one role: To explain how science achieved this unequalled status. For example, historians of science examined – and codified – scientific heroes. Philosophers of science explained how scientists produced certain knowledge of the natural world through the ingenious ways they integrated theoretical models and empirical evidence. Sociologists of science explained how the research community organized itself in a special way to produce its knowledge without contamination by outside influences, such as economic or political values that might have shaped or corrupted research. The first wave of scholarship essentially supported, without criticism, the scientific enterprise (Collins & Evans, 2002; Collins, 2014).

The second wave of science studies came in the 1960s. It dramatically changed the way scholars studied science and, as a consequence, came to view the nature of science and its place in society.

This second wave was catalyzed by Thomas Kuhn's *The Structure of Scientific Revolutions* (1962), which argued that science did not proceed in a linear fashion, with each new piece of knowledge seamlessly adding to existing knowledge, leading science progressively closer to the truths of the natural world. Instead, argued Kuhn, science developed through a series of revolutions, in which established theories, which could no longer explain new evidence that scientists produced, were replaced with new theories, or new paradigms, that gave fuller explanations of how the natural world worked. Kuhn in his book drew attention to one way in which science was communicated. He looked at the mini-histories of science in textbooks, which presented science as a story of linear progress produced by scientific heroes. "The potted histories at the start of textbooks are not," wrote Collins (2014, pp. 23–24), "serious history, but little fairy stories". And second-wave science studies, including much work by Collins and colleagues, sought to correct these fairy stories.

To do so, science studies scholars ventured into the laboratory to show the messy process of how scientists did their day-to-day work. These processes were hidden to non-scientists, but were understood by all scientists who undertook advanced research. These scientists, wrote Collins (2014, p. 24), "tend to get a nasty shock when they find that things aren't as straightforward as they had been taught and their job is really to organize a very untidy world – like trying to compress a balloon into a parcel tied with string." And social factors, second-wave scholars showed, were essential to scientific work – the social factors Collins observed at work among gravitational-wave physicists. Second wave scholars opened the inner workings of science to scrutiny, and as a result, levelled science, bringing scientists down from the exalted position they held for wave one scholars. Second-wave scholarship showed the external image of science – infallible heroes pursuing absolute truth – did not match the internal practice of science (Collins, 2014).

The third wave of science studies was first called for by Collins and Evans in 2002. They argued that second-wave scholars had destroyed the ideal image of science, stripping it of the myths and fairy stories that were a feature of wave one thinking. Third wave scholarship, in response, had to construct a realistic image of science. That realistic image must communicate that science is fallible and has limits and is carried out by people working in social groups, *while also* showing that science is still the optimal way for humans to understand an uncertain, complex natural world. Third wave scholars, argued Collins (2014, p. 81), must aim to "describe science accurately while still admiring it and what it stands for. The trick that has to be learned is to treat science as special without telling fairy stories about it."[2]

The rise of "technological populism"

But it has proved difficult for scientists to perform this trick. Many scientists, Collins argued, still communicate a first-wave image of science. They presented a heroic image of science as the pre-eminent form of knowledge, produced in isolation from wider culture. Especially when they seek to justify their work, scientists turned to what Collins (2014, p. 10) called science's "crown jewels" – the work of Newton or Einstein, for example, or esoteric, wondrous science, such as the search for the Higgs boson. Such first wave-anchored communication has created a problem for science's public image, because the science-related problems that citizens have faced are not ones that involved heroic science or abstract theorizing. The problems they faced featured contentious science, characterized by high uncertainty, where there were no clear-cut answers. As emblematic examples, Collins pointed to major food-related scandals in the United Kingdom in the 1990s and 2000s. The outbreak of bovine spongiform encephalopathy (BSE) in the 1990s was caused by feeding dead cattle to living cows as part of modern industrial farming processes. After telling people it was safe to eat beef, the UK government watched as more than 150 people died of BSE's human version, variant Creutzfeldt-Jakob disease (vCJD). Years later, a massive outbreak of another infection affecting cattle, foot-and-mouth disease, saw corpses of cows burned on pyres throughout the British countryside, as scientists argued about the proper response to the outbreak. "The public impression," wrote Collins (2014, p. 5), "was of an incompetent government, informed by feuding scientists, squandering our farming heritage and failing to handle another risk to the food chain".

Collins pointed to other areas of public life where expertise has proved contentious or unreliable. Despite decades of promise, medical science has failed to find cures for cancer and other serious diseases. As the 2009 financial crisis proved so devastatingly, economic modelling has been unreliable. Modern political movements around environmentalism and animal rights questioned expert knowledge, often with their own credentialed experts and bodies of knowledge (Collins & Evans, 2007). Moreover, Collins (2014) noted, there have been public concerns that science is manipulated by corporate funders, as exemplified by the public controversies over tobacco companies' influence on medical literature on smoking and oil companies' influence on the scientific literature on climate change. A consequence of these trends has been the rise of what Collins (2014, p. 15) called "default expertise". He argued that citizens have responded to the public exposure of the inherent uncertainties in science with "a sense of empowerment" that, because experts disagree intensely and expert predictions are fallible, their own views are as valid as those of scientists. "Possessing default expertise," wrote Collins (2014, 16), "means being as good as an expert because there are no experts." As Collins and Evans noted (2007, p. 2): "Our loss of confidence in experts and expertise seems poised to usher in an age of technological populism."

Their view was prescient. The notion of default expertise has become a characteristic of the contemporary era, which has been called an age of "post-factuality" or "post-truth." As defined by Oxford Dictionaries (2016), post-truth is an adjective that refers to "circumstances in which objective facts are less influential in shaping public opinion than appeals to emotion and personal belief." Bound up with post-truth is a mistrust of expert knowledge. In a notable and notorious example, in the run-up to the United Kingdom's referendum in 2016 about whether or not to leave the European Union (EU), the then justice secretary Michael Gove was asked in a media interview about economists' predictions about the impacts of leaving the EU. In response, he said: "people in this country have had enough of experts". He added: "I'm not asking the public to trust me. I'm asking them to trust themselves" (cited in Mance, 2016).

Gove's quote illustrates what Collins and Evans have long argued is the fundamental tension in the contemporary relationship between science and society: The tension between expertise and democracy. As they noted, in contemporary culture, the public has the political right to contribute to debates over science and technology. Without public contribution, scientific developments are likely, at worst, to be opposed, or, at best, are likely to be not fully trusted by citizens. Collins and Evans (2002, p. 235) have called this the "problem of legitimacy." But this leads to a related problem. What are the legitimate contributions that non-experts can make to the technical dimensions of scientific debates? How can boundaries be set around these public contributions? What is the boundary of a scientist's expertise? How far does a scientist's expertise extend? Collins and Evans (2002, p. 235) have called this the "problem of extension." Scientists who communicate must address these problems. And trainers have a role in making sure scientists do this explicitly, by laying out a clear understanding of different varieties of expertise, which scientists can use to underpin their communications work. And expertise is central to science communication. As one recent review of training programs argued (Baram-Tsabari & Lewenstein, 2017, p. 288): "Good science communication enables its audience to make informed judgments about sources and expertise, regardless of an audience's technical knowledge."

A taxonomy of expertise

Collins and Evans (2007, p. 14) set out a clear way to understand expertise. They contrasted default expertise with substantive expertise, which they categorized into what they called a taxonomy of expertise, or "the periodic table of expertises". The major types of substantive expertise in their typology are:

1 *Ubiquitous Expertise*. This is the expertise everyone in a culture acquires, such as speaking a native language and knowing unwritten social rules, such as table manners and personal space. It does not require effort to acquire this expertise, as it is assimilated as a result of growing up in a particular society.
2 *Specialist Expertise*. This is the form of expertise most associated with scientists. It is the type of expertise possessed by someone who has engaged

in a prolonged period of practice to develop specialist knowledge. It is the expertise possessed, for example, by a professional chemist, mathematician, engineer, truck driver, carpenter, or violinist. When applied to scientists, in particular, there are two sub-components of specialist expertise. The first, for Collins and Evans (2007, p. 35), is "interactional expertise," which is a level of almost fluent knowledge acquired through the process of interacting with an expert community, but without contributing to the activities of that community. The second, for Collins and Evans (2007, p. 24), is "contributory expertise," which is acquired by contributing original knowledge to a specialist field, such as new data, concepts, or theories in a professional field. Contributing experts are all interactional experts, but not all interactional experts are contributing experts.

3 *Meta-Expertise*. This is a high-level form of expertise in which someone has acquired the ability to, in Collins's words (2014, p. 59), "judge and choose between other experts; in principle, this kind of expertise can be good enough to guide one through the decisions one has to make in the contemporary technological world". Collins and Evans (2007) provided three criteria used by someone with meta-expertise to evaluate experts: Credentials, experience, and track record. Of the three, track record is the best criterion. However, this form of meta-expertise is difficult to acquire at a high level, and is usually developed by those who have studied science in close and immersive detail for a long time, such as experienced science journalists.[3]

Insights for communication training

Collins and Evans persuasively argued in their work that scientists' legitimacy in public life is based on their specialist expertise and experience, as well as their adherence to the scientific ethos. Their expertise and ethos have set them apart from other experts. Their expertise, argued Collins and Evans (2002, p. 236), "is the reason for using the advice of scientists and technologists in virtue of the things they do *as* scientists and technologists" compared to what other experts do in their respective roles. Based on their scholarship, I have inductively drawn out five insights or themes for communication training, which trainers can potentially incorporate into the design of their programs, and which scholars can investigate in their research.

1. Scientists should communicate only from their core area of expertise. Scientists, Collins and Evans (2007) noted, cannot speak with much authority, if any, outside their specialist fields. In fact, the further researchers stray from their expert domain, the closer their communication comes to the fairy stories about science that were a feature of first wave science studies. When scientists speak outside their specialization, they are not providing expert interpretation. They are, for Collins and Evans (2007, p. 145), providing "punditry." Speaking outside their expertise reduces their authority, raising the problem of legitimacy. This insight is fundamental to their scholarship and it has clear implications for scientists who seek in training the skills to become defenders of the entire scientific enterprise. Collins and Evans wrote (2002, p. 270), "if there are to be public defenses of science, they should concentrate on scientists as specialists, rather than as generalists".

2. Scientists must continuously grapple with the problem of extension. When scientists communicate in public, they will undertake roles as "public experts" (Peters, 2014, p. 70). In this role, scientists are often asked, usually by journalists, to provide concrete advice to citizens and decision-makers about specific situations related to their area of expertise. The challenge for scientists in this role is that they will be expected to offer expert knowledge that helps citizens or decision-makers decide on future courses of action. They also face the challenging situation of being asked to comment on social or political impacts of science, areas that are situated at the very edge of their expertise. As Peters noted:

> Scientists who present themselves as public experts are responsible, first, for acquiring and using the full available knowledge relevant to the problem; second, for making a systematic and comprehensive assessment; and, third, for communicating it in a way that supports the decision-making of members of the public.
>
> (Peters, 2014, p. 73)

It follows from the work of Collins and Evans that scientists can only undertake this role if they are communicating their particular expert knowledge. Only by drawing on their expert knowledge can they undertake the three responsibilities outlined by Peters, responsibilities that involve scientists continuously negotiating the problem of how far their expertise extends. Trainers will, therefore, need to focus on helping scientists think through this problem of extension as it applies in their communication endeavors.

3. Scientists should convey the nature of their expertise. When communicating their science, researchers should also communicate their expertise. In this communication, scientists should explain how they developed their interactional and contributory expertise. Not only does this provide a basis for their own authority, but it also conveys the social dimensions of knowledge production. As Collins and Evans (2007, p. 140) argued, those who communicate science must "be ready to explain science, and explain the nature of expertise, to as wide an audience as possible."

4. The scientific ethos should be conveyed. Connected to conveying the nature of expertise, scientists should communicate how they adhere to the scientific ethos in their work. As Collins (2014) argued, the scientific ethos is a distinctive part of the culture of science, and so the values that drive their individual and collective work is an essential point for non-specialists to understand. This involves the explanation of how norms including disinterestedness, skepticism, and universalism have played out in the creation of expert knowledge around a particular topic. Carefully communicating this process can vividly illustrate a central aim of Collins and Evans's work: Although an enterprise produced by fallible humans, science is still the best way we have for understanding the natural world. Understanding science means understanding the scientific ethos.

5. Scientists must acknowledge their cultural context. In the contemporary *zeitgeist* of the west, expertise is challenged, knowledge is contentious, and scientific findings – especially as applied to social problems – are uncertain. This is the cultural background in which scientists are communicating. While many training programs emphasize tailoring messages to resonate with audience values, the wider cultural context is often neglected. Scientists, with the help of communication trainers, should understand and address their cultural context, as it will impact on how their ideas are received by members of different publics. "As a general rule," wrote the political communications scholar Brian McNair (2011, p. 29),

> the effects of political communications of whatever kind are determined not by the content of the message alone, or even primarily, but by the historical context in which they appear, and especially the political environment prevailing at any given time.

Conclusion: training scientists to communicate expertise and experience

The ideas of Collins and Evans offer a conceptual foundation for making sense of the role of scientists in contemporary culture and society. Their science studies scholarship demonstrates that expertise and experience are at the core of the unique work scientists do and, by extension, what scientists communicate about. Collins and Evans present a strong case that scientists must address in their public communication the problem of legitimacy and the problem of extension. The way to address these connected problems is to focus on expertise and experience. Paradoxical as it seems, scientists must argue that they have a legitimate place in public discussion about science-related problems. They must do so by conveying the nature of their particular expertise, communicating how that expertise was developed, both through interactions with other experts and through contributing original knowledge to their area of expertise. Through the communication of their specialist expertise, scientists will be in a position to convey the core elements of the scientific ethos, the values that distinguish scientists and scientific work. In an age of post-truth and default expertise, where the scientific issues that most affect citizens are uncertain and contentious, scientists must be clear about how far their expertise extends. Communication trainers have a significant role to help scientists think this point through. The work of Collins and Evans, furthermore, provides a conceptual framework for creating training programs that have at their core the communication of expertise.

Their ideas are useful also for communication training scholars. The rapid expansion of training as a field of practice and research has meant the sub-field of scholarship has been marked by a theoretical incoherence and a lack of foundational ideas (Baram-Tsabari & Lewenstein, 2017). As expertise and experience are fundamental to science, expertise and experience are fundamental to science communication. Scholars of communication training can make the studies of expertise and experience central to their research, as they explain theoretically how scientists address the problems of legitimacy and extension in various communication contexts. By providing clear conceptual accounts of how scientists convey their expertise and experience and ethos, scholars can provide a clear account of the role of science in contemporary culture.

Bringing expertise to the core of communication training also has democratic implications. Citizens face the ongoing challenge of making good democratic decisions about scientific issues which they do not have the specialist expertise to understand fully. When citizens make these judgments, write Collins and Evans (2007, p. 139), they do so by "choosing *who* to believe rather than *what* to believe." They are evaluating the experts. Communication trainers have a role in helping citizens make good decisions. They can do this indirectly, by training scientists to explain the social process through which they came to possess their specialist technical expertise. Citizens, as a consequence, will have more knowledge on which to base their evaluation of experts. As Collins and Evans concluded (2007, p. 139): "If those quasi-technical judgments open to the technically inexperienced citizen are to be based on social judgments, then the better the understanding of the social processes of science, the better the judgments are likely to be."

Notes

1 The ESConet modules are available as a pack under creative commons at: https://esconet.wordpress.com/training-materials. The quotations about the objectives of Science in Culture are taken from p. 113 and p. 115 of the module pack.

2 It is not that Collins and Evans's work is without flaws. Their conceptualization of three waves of science studies has been the subject of much intense debate and criticism (see Wynne, 2002 and Jasanoff, 2003). And personally, I do not agree with the relativist and social constructionist dimensions of their work, as I hold a critical realistic epistemological position. However, even with these criticisms, the work of Collins and Evans has much to offer science communication training.

3 The description of the typology presented here is an overview of a more detailed typology developed in chapters two and three of Collins and Evans (2007).

References

Baram-Tsabari, A., & Lewenstein, B. V. (2017). Science Communication Training: What Are We Trying to Teach? *International Journal of Science Education*, *7*(3), 285–300.

Besley, J. C, Dudo, A. D., Yuan, D., & Abi Ghannam, N. (2016). Qualitative Interviews With Science Communication Trainers About Communication Objectives and Goals. *Science Communication*, *38*(3), 356–381.

Collins, H. (2014). *Are We All Scientific Experts Now?* Cambridge: Polity Press.

Collins, H. & Evans, R. (2009). *Rethinking Expertise*. Chicago: University of Chicago Press.

Collins, H. M. & Evans, R. (2002). The Third Wave of Science Studies: Studies of Expertise and Experience. *Social Studies of Science, 32*(2): 235–296.

Crow, M. & Sarewitz, D. (2013). Power and Persistence in the Politics of Science. In G. P. Zachary (Ed.) *The Rightful Place of Science: Politics*, 1–8. Tempe, AZ: Consortium for Science, Policy and Outcomes.

Dudo, A. & Besley, C. (2016). Scientists' Prioritization of Communication Objectives for Public Engagement. *PloS ONE, 11*(2). Retrieved from e0148867.doi:10.1371/journal.pone.0148867

ENSCOT. (2003). ENSCOT: The European Network of Science Communication Teachers. *Public Understanding of Science*, *12*(2), 167–181.

Fahy, D. (2015). *The New Celebrity Scientists: Out of the lab and Into the Limelight.* Lanham, MD and London: Rowman & Littlefield

Giles, J. (2006). Sociologist Fools Physics Judges. *Nature*, *442*, 8.

Goodell, R. (1977). *The Visible Scientists.* Boston, MA: Little, Brown & Company.

Jasanoff, S. (2003). Breaking the Waves in Science Studies: Comment on H. M Collins and Robert Evans, 'The Third Wave of Science Studies' *Social Studies of Science*, *33*(3), 389–400.

Kuhn, T. (1962). *The Structure of Scientific Revolutions.* Chicago: Chicago University Press.

McNair, B. (2011). *An Introduction to Political Communication* (5th edition). Abingdon: Routledge.

Mance, H. (2016). Britain has had enough of experts, says Gove. *Financial Times.* Retrieved from www.ft.com/content/3be49734-29cb-11e6-83e4-abc22d5d108c

Merton, R. K. (1973). *The Sociology of Science: Theoretical and Empirical Investigations.* Chicago: University of Chicago Press.

Miller, S., Fahy, D., & the ESConet Team. (2009). Can science communication workshops train scientists for reflexive public engagement? The ESConet experience. *Science Communication*, *31*(1), 116–126.

Neeley, K. A. (2010). Science, Technology, and Society Studies. In S. H. Priest, *Encyclopedia of Science and Technology Communication*, (pp. 737–742). Los Angeles: Sage.

Oxford Dictionaries. (2016). Word of the Year 2016 is ... *Oxforddictionaries.com.* Retrieved from: https://en.oxforddictionaries.com/word-of-the-year/word-of-the-year-2016

Peters, H. (2014). Scientists as Public Experts: Expectations and Responsibilities. In M. Bucchi & B. Trench (Eds.), *The Routledge Handbook of Public Communication of Science and Technology* (pp. 70–82). Abingdon: Routledge.

Pielke, Jr., R. (2007). *The Honest Broker: Making Sense of Science in Policy and Politics.* Cambridge: Cambridge University Press.

Pinker, S. (2019). *Enlightenment Now: The Case for Reason, Science, Humanism, and Progress.* Penguin: London.

Sheehan, H. (2007). Marxism and Science Studies: A Sweep Through the Decades. *International Studies in the Philosophy of Science 21*(2): 197–210.

Wynne, B. (2002). Seasick on The Third Wave? Subverting the Hegemony of Propositionalism: Response to Collins & Evans (2002). *Social Studies of Science 33*(3): 401–417.

Ziman, J. M. (1991). *Reliable Knowledge: An Exploration of the Grounds for Belief in Science.* Cambridge: Cambridge University Press.

3 The meaning of public–private partnerships for science communication research and practice

Fred Balvert

Introduction

Current perspectives on science communication, including Responsible Research & Innovation (RRI), the science in society framework of the European Union, promote engagement of all societal stakeholders in all stages of scientific research. These stakeholders include industrial partners – companies that pursue economic profit. But the role of companies in research is problematic to science communication. Historically, it does not fit into the discourse of science communication as a discipline that has developed since the second half of the twentieth century. In order to stay relevant, science communicators have to develop meaningful modes of dealing with industrial partners in science communication theory as well as in practice.

Background

Since the Second World War, science has been developing and booming as a public enterprise. During and after the war it had become clear to governments that national interests, such as security, energy, health, infrastructure, education and economic development depend largely on scientific progress. For the first time in history national governments perceived it as their task to interfere with science. In order to do so, modern national science policies were defined. "The twentieth century has witnessed the evolution of the practice of science and technology from a predominantly individual investigator, low-budget, privately financed mode to multidisciplinary, high-budget, publicly financed research teams." (Miller, 1983).

This has strongly influenced the development of science communication. One of the main reasons for communicating science was to promote public support for government spending on research. For example, the first scientific literacy programs of the US Office of Education and the National Science Foundation were undertaken primarily to move students towards scientific careers and to engender public support for the costs and risks of cold war science. (Paisley, 1998). In 1960 the National Science Foundation started a campaign for the Public Understanding of Science, aimed at securing broad support among the public for the financing of science and technology. (Wiedenhoff, 2000).

This foundation of science communication is still very present in theory and practice today. The textbook *Successful Science Communication: Telling it like it is* reports on a public dialogue about nanotechnology for healthcare: "There were concerns about who benefits from the expenditure of public funds on science." (Jones, 2011).

Appeasing values

National science policies of modern democratic countries and consequently the field of science communication itself, combine important, but distinctive values of scientific culture and of democratic policy making. Scientific culture is striving for excellence, assessed through the peer review process, while democratic policymaking is striving for societal relevance, which is assessed by open and transparent evaluations. A model of appeasement of both value systems was first described by Vannevar Bush in his report "Science, the endless frontier" (Bush, 1945) and has been summarized by Jasanoff: "In exchange for continued governmental support and freedom to define their research priorities and methods, scientists would provide the public with beneficial discoveries and a trained workforce." (Jasanoff, 2005).

Much has changed in science policy since the early days of post-war modernism. Presently, national and European science policies are actively promoting public–private collaboration in research and innovation. In addition to the values of scientific culture and democratic policymaking, these policies have introduced a third value system, that of the market. As a result, public funding of research is no longer the only and most important way of funding research that demands legitimization. Another issue that has emerged is public support for the funding of research by private parties and industries.

Science and the market

During the last decades, radical changes have been taking place in the science policies of most national governments and the European Union. One is the introduction of market economic principles. This can be considered as an outcome of the neo-liberal movement in public administration, known as New Public Management, that has changed the way in which public services are being financed. It has had profound effects on scientific research (Elzinga, 2010).

The public–private partnership has become a widely accepted organization model in which "governments are offering incentives, on the one hand, and pressing academic institutions, on the other, to go beyond performing the traditional functions of cultural memory, education and research, and make a more direct contribution to 'wealth creation'." Etzkowitz and Leydesdorff point out that:

> [A] new social contract between the university and the larger society is being negotiated in much more specific terms than the old one. The former

> contract was based on a linear model of innovation, presuming only long-term contributions of academic knowledge to the economy. Now, both long- and short-term contributions are seen to be possible, based on examples of firm formation and research contracts in fields such as biotechnology and computer science. A spiral model of innovation is required to capture multiple reciprocal linkages at different stages of capitalization of knowledge.
> (Etzkowitz & Leydesdorff, 1995)

This spiral is known as the "triple helix model," which implies that the government, academic institutions and private companies constitute the three partners that collaborate in the organization of scientific research, each contributing its own unique qualities. And more recently, the "quadruple helix" has been introduced, which also involves the end-users of research-driven innovation, such as consumers and patients (Carayannis & Campbell, 2009).

The triple helix and quadruple helix have become the normative models which are applied in the procedures of research funding of national and European science policies. It has resulted in new criteria for publicly funded research. Researchers have to convince funding organizations that their research is contributing to economic competitiveness and have to seek public–private partnerships, meaning that private parties are participating and partly funding research. "Scientific knowledge has become a commodity: it is the basic raw material of our age." (Gregory, 2016)

Current science policies in Europe, Brazil and the Netherlands

An analysis of current public science policies in the European Union, the Brazilian state of São Paulo and the Netherlands shows a wide consensus about the modernist match between science and society as well as a shift towards the utilization of science for economic policy. The three current policies all reflect the logic of securing space for freedom of inquiry in return for the benefits that science delivers to society. But a new element compared to former iterations of these policies is that they each state that science should be seen as a driver of innovation, economic prosperity and employment.

Of course, this shift in thought gives direction to the criteria for the distribution of research budgets. The largest part of these budgets that the three policies distribute is reserved for problem-oriented research based on two major criteria: Contributions to the solution of global societal problems and to innovations which can bring about economic advantages.

European Union

Horizon2020 is the science policy framework of the European Union for research from 2014 to 2020, with a total budget of €79 billion. Its mission states that it:

> [It] is the financial instrument implementing the Innovation Union [...] aimed at securing Europe's global competitiveness. Seen as a means to drive economic growth and create jobs, Horizon2020 has the political backing of Europe's leaders and the members of the European parliament. They agreed that research is an investment in our future and so put it at the heart of the EU's blueprint for a smart, sustainable and inclusive growth and jobs.
>
> (European Commission, 2016)

Most of the budget is distributed over three work programs: Excellent science (32%); Societal challenges (39%); and Industrial leadership (22%). In all three work programs researchers are stimulated to seek collaboration with companies in public–private partnerships.

The Brazilian State of São Paulo

Fapesp (Fundação de Amparo à Pesquisa do Estado de São Paulo), the São Paulo Research Foundation, 'is a public foundation, funded by the taxpayer in the State of São Paulo, with the mission to "support research projects in higher education and research institutions." (Fapesp, 2016). "Fapesp believes that advancing human knowledge generates benefits for the progress of humanity." (Fapesp, 2013). In 2012, Fapesp spent the equivalent of $530 million on research based on its three main criteria: Advancing knowledge; Research for practical application; Research infrastructure. Within this breakdown "Advancing Knowledge" represents the classical aim of training human resources and fostering academic research to which 37% of expenditures has been allocated. The line of "Application-Driven" research is directed to serving economic and societal interests to which 53% has been allocated. The remaining 10% of the budget has been spent on research infrastructure.

The Netherlands

The Netherlands Organization for Scientific Research (NWO) "funds top researchers, steers the course of Dutch science by means of research programs and by managing the national knowledge infrastructure." (NWO, 2010). Its yearly budget of around 683 million euro (2014) is distributed over policy instruments, ranging from national research grants to subsidies for research in accordance with the "top sectors," national strategic goals for research and development related to knowledge-driven economic sectors.

The breakdown of the NWO budget combines curiosity-driven research with politically defined objectives. Among the goals stated in the strategic plan 2011–2014 are: 1. Invest in talent and free research; 2. Invest in society-inspired themes in collaboration with partners; 3. Stimulate and facilitate the application of knowledge. Although in 2013, 45% of the budget of €628 million was invested in talent and free research (1), this research was in fact closely linked to society-inspired research within public–private partnerships (2), and technology transfer (3). In 2015, on top of this distribution, an estimated amount of €40 to €85 million for free research was used for research related to the top sectors.

Public trust

From a public perspective the interference of private parties in research is controversial. The influential report by the UK House of Lords (2000) pointed out that "survey data reveal [...] negative responses to science associated with government or industry, and to science whose purpose is not obviously beneficial. These negative responses are expressed as a lack of trust." These findings are supported by those of academics. Millstone and Van Zwanenberg write that:

> [A]mong many groups, there has been a decline in the levels of trust in particular groups of scientists, such as those working in, or for, the companies and industries whose products and processes are under scrutiny. Research also indicates that the levels of trust in scientists working in, or for the government are very low.
>
> (Millstone & Van Zwanenberg, 2000)

Clearly, this affects trust in science in general, especially since the mentioned scientists working in, or for companies, industries and the government often hold positions at universities, where contracted research has become normal.

Distrust of researchers who associate themselves with companies or industrial partners is also common among professionals who are well informed about scientific practice, and among scientists themselves. A survey among a mixed group of forty researchers, science journalists and science communicators participating in the summer course of Erice International School of Science Journalism in 2016 reveals that trust in the results of research decline if researchers are working together with the industry.

91% of respondents scored the estimated reliability of research results reported by scientists in general as 6 to 10 on a 10-point scale, 10 being very reliable. This number dropped to 63% when scientists are working directly for a company, and to 68% when scientists working at a university are contracted by companies.

This estimated reliability of researchers was lower after a four-day program of lectures about science, science journalism and science communication in which various aspects, advantages as well as concerns, of public–private collaboration were presented and discussed: The scores respectively were 86%, 54% and 56% (although these scores may have been affected by the fact that 38 respondents completed the first questionnaire and only 34 completed the second).

Criticism related to funding also echoes within the scientific community. In his article "Why Most Published Research Findings Are False," John Ioannidis points out that the greater the financial interests in a scientific field, the less likely the research findings are to be true (Ioannidis, 2005).

Table 3.1 Reliability of reported research results estimated by fellows of Erice International School of Science Journalism (summer course, 2016) in percentages of respondents

		Very unreliable ←									*→ Very reliable*
		1	*2*	*3*	*4*	*5*	*6*	*7*	*8*	*9*	*10*
How reliable, would you say, are research results reported by scientists in general?	Pre (n=38)				5	5	11	24	45	8	3
	Post (n=34)			6	3	6	6	15	50	15	
How reliable, would you say, are research results reported by scientists who are working for a company?	Pre (n=38)			11	5	21	26	18	8	8	3
	Post (n=34)		6	15	9	18	30	21	3		
How reliable, would you say, are research results reported by scientists working at a university, whose research is commissioned by companies?	Pre (n=38)		3	8	3	18	18	26	13	11	
	Post (n=34)		6	9	15	15	26	18	12		

Call for theory

As we have seen, it has been the initial mission of science communication to legitimize the public funding of science. Presently, the public funding of research is no longer the only way of financing that demands legitimization. Another one is the funding of research by private parties and industries, since public–private partnerships have become common practice and are promoted by public science policies and research institutes alike. But in science communication theory the profound effects of the introduction of marketing economic principles in research still seem underestimated.

Science communication literature until now has shown little interest in this game changer. “The dominant institutional patterns as well as powerful economic and political imperatives, such as enhancing national competitiveness, economic growth and techno-scientific innovation, remain unchallenged.” (Braun & Könninger, 2018).

Various authors have paid attention to industrial parties as stakeholders with their own specific characteristics. For example, in a neutral way as “actors in the governance of science” (Bandelli & Konijn, 2012), as parties related to a decline in trust (Millstone & Van Zwanenberg, 2000) or as outright ‘villains’ (Wagner-Egger et al., 2011). But few articles have been published about the effects of the increasingly prominent role of public–private collaboration in research and the implications for science communication.

There are some exceptions, such as the article “Understanding the impact of commercialization on public support for scientific research: Is it about the funding source or the organization conducting the research?" published in the journal *Public Understanding of Science*. The authors confirm "that support drops significantly when scientific research is funded by private rather than public interests, and even more so when it is conducted in a private company rather than a public university." (Critchley & Nicol, 2011). They call for future research that illuminates "factors that are directly associated with different research organizations and funding sources." Weingart and Guenther started a discussion in their article in *JCOM* (2016). In a responding article Irwin and Horst state that "rather than presenting industrial research as a deviant form of science, we need to maintain a critical, open, and empirically-based perspective on the activities in question" (2016).

A critical attitude towards interference of companies with research is necessary considering the dependent position of researchers on funding and the primary aim of companies to make profit. The embrace of companies as stakeholders by science policy makes it unavoidable for science communication to chart this terra incognita.

How could practice respond?

To keep up with the changing reality of the funding of scientific research, while acknowledging the negative connotation of public–private collaboration in research, although science communication research should clearly pay more attention to this important aspect of scientific practice, there is obviously also a task for science communication practitioners in this area.

Generally speaking, addressing ethical and controversial issues should be a strand of science communication practice. The public is entitled to have the information it needs to perceive and assess research in its true and proper context in order to form an opinion about it. To promote this, science communication should not only focus on scientific concepts and results, but should also explain the practice of scientific conduct in an open and transparent way. Besides, my own experiences with communication about animal research at Erasmus MC, for example, are that communicative interventions are effective in changing public attitudes and countering prejudices. This does not rule out a critical dialogue; on the contrary it is often when a dialogue starts. Contrarily, not addressing controversial aspects of scientific practice will undermine public trust in science on the long run.

The scientist as a salesperson

Effective science communication should start at the heart of research itself. A European survey showed that a majority of citizens (66%) feel that scientists working for a university (not politicians, government representatives or journalists) are the best qualified people to explain the impact of scientific and technological developments on society (European Commission, 2013).

At the same time, scientists have to prove that their research contributes to solving societal problems and to economic activity, while external financers, whether these are governments or private parties, have an interest in the visibility of the research and thereby indirectly of themselves (Balvert, Hulskamp, & Zgaoui, 2014, p. 19). If it is the task of science communication to contribute to the transparency of funding, this task starts with the researchers and research institutes, which constitute the procedures of scientific practice.

The position of researchers today is defined by an interplay of the different interests that are at stake, which reflect the dimensions of the quadruple helix: The interest of research institute is academic prestige; the interest of governmental funding institutes is societal relevance; the interest of industry is commercial advantage, and the interest of the consumer is ensuring a maximum of practical benefit at minimum or at least acceptable costs and risks. These interests do not necessarily coincide, in particular the collaboration between research, which should in all circumstances be independent, and private companies pursuing profits, could lead to a (perceived) conflict of interest. Although research that has been contracted by government agencies for the purpose of science-based policy, is regularly under scrutiny as well. Unfortunately, ties between researchers and the industry lie at the base of many cases of scientific misconduct, controversies and public scandals.

This is why the position of the researchers, and consequently that of science communicators, has not become easier with the worldwide introduction of market economic criteria into science policies. In order to avoid perceived or actual conflicts of interest, scientific integrity and transparency should be safeguarded from the beginning of every research project. For the researcher, this means (s)he should be able to explain that legal and ethical criteria have been taken into account and are met in all stages and all aspects of the project. Although messages directed to different stakeholders and audiences involved may vary because of their different information needs, these messages should never be contradictory or incomplete, to avoid situations where integrity could be called into question. Every decision to withhold information from any specific audience implies the risk that the researcher and the research itself will be perceived or proven to be unreliable.

Clearly, the responsibility for this lies not with the researcher alone, but with all stakeholders involved, especially with the research institute where the researcher is working. Unambiguous institutional policies should be in place concerning scientific integrity, transparency, funding and disclosure of interests (DoI), which are consistent with national policies and legislation. Those policies should be proactively communicated and discussed within the research community of the institute. Most research institutes have introduced Technology Transfer Offices (TTO) and frameworks for valorization to facilitate the shift toward public–private collaboration and to help researchers in combining their academic, societal and entrepreneurial roles.

For science communicators, either internal or external to the research institute, national and institutional policies and regulations, together with broader societal and ethical concerns, constitute the context of their work. It is their role to proactively and reactively enlighten the respective audiences about the "social contract" under which research is being done.

A good lesson learned from practice is that communication aspects of research projects in which companies are involved should be the topic of explicit arrangements in the contract between the parties, worked out in a communication paragraph that could reference a communication plan. During the project different views may evolve on the questions why, how, when and by whom to communicate the progress or the results of the research. For example, think of a clinical study in which patients of minor age are involved, where the pressure on the industrial partner to publish preliminary results in order to boost stock value could easily overrule the interests of the other stakeholders – researchers, physicians, patients and parents – if arrangements on publicity are not part of a legally binding contract. In general, it is much easier to make these agreements before the stakes get too high. Institutional science communicators are (or should be) in a position to advise scientists, the board and TTO on this.

It is a responsibility of science communicators, whether they are working within research institutes as communication advisors or press officers, or outside as science journalists, organizers of events or curators of science museums, to emphasize the need for transparency, to facilitate and promote it and to explain the constraints and consequences of science policies and practices.

Towards models for practice

Several models are worth exploring to make sense of the relations between researchers and private enterprise in the current practice of scientific research.

Watchdog model

In this model science communicators develop a critical stance on the role that private parties are supposed to play in research. Keeping in mind the problematic implications of the dynamics between the values of scientific excellence, societal relevance and economic competitiveness, the relations between researchers and private partners can be investigated and presented to the public. Examples of communication modes that suit this model are journalistic productions, debates, blogs, videos, books.

Informative model

Explaining the role of private parties in the continuum from fundamental research and innovation, to application and market valorization is a positivist narrative that sheds light on current scientific practice in the wider context of economic growth and employment. Based on the premise that the public is very well able to understand examples of public–private collaboration, it is the role of the science communicator to be critical towards interests and expectations of all stakeholders during the process and to address these adequately in modes of representation. This model can be applied through many communication modes, such as exhibitions, festivals, printed/online publications and games.

Collaborative model

Given the role of companies in research projects and the task of science communicators to communicate about these projects, for example because they are working for one of the project partners (academic, government, industrial or NGO), science communication itself can become a part of public–private collaboration. In such cases, it is important that independence in the creation of content by the science communicator is negotiated and safeguarded right at the start of the project, for example by installing an independent advisory board. A Disclosure of Interests should be brought effectively to the attention of the public, e.g., as part of the colophon in which all contributors of the project are mentioned. This model can be applied to the same wide range of communication modes, such as exhibitions, festivals, printed/online publications and games.

Entrepreneurial model

Science communication can also serve the commercial interests of private companies or the industry. This is the case when science communication projects are sponsored or commissioned by a private company, or when a private company is actually organizing the science communication project. In the case of sponsorship, a contract should be in place that rules out any influence by the sponsor on the content of the project. Again, a Disclosure of Interests, for example in the colophon of the project, is a necessity. In this model the responsibility of the science communicator to pay respect to the distinctive roles and interests of the stakeholders remains of primary importance. Obscuring commercial interests is not only unethical but also jeopardizes the legitimacy and perception of the project and of scientific research in general. By the same token, addressing and explaining the context of public–private collaboration will contribute to public understanding and support. All modes of communication mentioned above are suitable, but will have a more commercial character in the form of sponsorship or advertising.

References

Balvert, F., Hulskamp, M., & Zgaoui, S. (2014). *Prepare for 15 seconds of fame: Media contacts for researchers*. Rotterdam: Trichis Publishers.

Bandelli, A., & Konijn, E. A. (2012). Science Centers and Public Participation: Methods, Strategies, and Barriers. *Science Communication*, *35*(4), 419–448. doi: 10.1177/1075547012458910

Braun, K., & Könninger, S. (2018). From experiments to ecosystems? Reviewing public participation, scientific governance and the systemic turn. *Public Understanding of Science*, *27*(6) 674–689. doi: 10.1177/0963662517717375

Bush, V. (1945). *Science, The Endless Frontier.* Washington: National Science Foundation.

Carayannis, E. G. & Campbell, D. F. (2009). 'Mode 3' and 'Quadruple Helix': toward a 21st century fractal innovation ecosystem. *International Journal of Technology Management, 46*(3–4), 201–234. doi: 10.1504/IJTM.2009.023374

Critchley, C. R., & Nicol, D. (2011). Understanding the impact of commercialization on public support for scientific research: Is it about the funding source or the organization conducting the research? *Public Understanding of Science, 20*(3), 347–366. doi: 10.1177/0963662509346910

Elzinga, A. (2010). New Public Management: Science policy and the orchestration of university research– academic science the loser. *The Journal for Transdisciplinary Research in Southern Africa, 6*(2), 307–332. Retrieved from http://dspace.nwu.ac.za/handle/10394/3861

Etzkowitz, H. & Leydesdorff, L. (1995). The Triple Helix–University–Industry–Government Relations: A Laboratory for Knowledge Based Economic Development. *EASST Review 14*(1), 14–19. Retrieved from http://dare.uva.nl/document/2/935

European Commission. (2013). Special Eurobarometer 401 on Responsible Research an Innovation (RRI).

European Commission. (2016). Horizon2020 website: https://ec.europa.eu/programmes/horizon2020/en/what-horizon-2020

Fapesp. (2013). Fapesp 2012: Annual Activity Report. São Paulo: São Paulo Research Foundation. Retrieved from www.fapesp.br/en/5437

Fapesp. (2016). Fapesp website. www.fapesp.br/en/5385

Gregory, J. (2016). '"The price of trust—a response to Weingart and Guenther". *JCOM, 15*(06), Y01.

House of Lords. (2000). Third report: Science and society. Retrieved from www.publications.parliament.uk/pa/ld199900/ldselect/ldsctech/38/3801.htm

Ioannidis J. P. A. (2005). Why Most Published Research Findings Are False. *PLoS Med, 2*(8), e124. https://doi.org/10.1371/journal.pmed.0020124

Irwin, A. & Horst, M. (2016). Communicating trust and trusting science communication—some critical remarks. *JCOM, 15*(06), L01.

Jasanoff, S. (2005). Judgement under siege: The three-body problem of the expert legitimacy. In S. Maasen & P. Weingart (Eds.), *Democratization of Expertise? Exploring novel forms of scientific advice in political decision making. Sociology of the Sciences,* (pp. 209–224). Retrieved from http://link.springer.com/chapter/10.1007%2F1-4020-3754-6_12

Jones, R. J. (2011). Introduction: Public engagement in an evolving science policy landscape. In D. J. Bennet & R. J. Jennings (Eds.), *Successful Science Communication: Telling It Like It Is* (pp. 1–13). Cambridge: Cambridge University Press.

Miller, J. D. (1983). *The American People and Science Policy: The role of public attitudes in the policy process.* New York: Pergamon Press.

Millstone, E. & Van Zwanenberg, P. (2000). A crisis of trust: For science, scientists or for institutions? *Nature Medicine, 6,* 1307–1308. doi: 10.1038/82102

NWO. (2010). Nederlandse Organisatie voor Wetenschappelijk Onderzoek. *Groeien met kennis: Strategienota NWO 2011-2014.* Den Haag: NWO. Retrieved from www.nwo.nl/overnwo/X+publicatie/nwo/strategienota-2011-2014-groeien-met-kennis.html

Paisley, W. J. (1998). Scientific literacy and the competition for public attention and understanding. *Science Communication, 20*(1), 70–80. doi: 10.1177/1075547098020001009

Wagner-Egger, P., Bangerter, A., Gilles, I., Green, E., Rigaud, D., Krings, F., … Clémence, A. (2011). Lay perceptions of collectives at the outbreak of the H1N1 epidemic: heroes, villains and victims. *Public Understanding of Science*, *20*(4), 461–476. doi: 10.1177/0963662510393605

Weingart, P., & Guenther, L. (2016). Science communication and the issue of trust. *JCOM*, *15*(5), C01.

Wiedenhoff, N. (2000). Wetenschaps- en techniekvoorlichting: Op zoek naar balans tussen Apollo en Dionysos. *Gewina*, *23*, 228–239. Retrieved from: www.gewina.nl

4 Science engagement and social media

Communicating across interests, goals, and platforms

Emily Howell and Dominique Brossard

The information and news environment is increasingly composed of online media, especially through the constant growth and evolution of social media platforms since the early 2000s (Greenwood, Perrin, & Duggan, 2016; Newman et al., 2017). This shift is most pronounced for science information and news in particular. As legacy newspapers shrank, they often cut their science sections before other news beats (Brossard, 2013; Brossard & Scheufele, 2013; Brumfiel, 2009; Newman et al., 2017; Peters et al., 2014), and science journalists, communicators, interested publics, and scientists themselves are migrating to online-only mediums, such as blogs and social media platforms, to share and learn science-related information (Brumfiel, 2009; Brossard & Scheufele, 2013; Pew Research Center, 2016).

Because of these changes, scientists and science communicators increasingly rely on social media to engage with peers, stakeholders, and interested publics. The new and changing social media environment, however, also comes with features that can facilitate or limit successful communication and engagement. As bad communication is often worse than no communication, it is important for communicators to understand the features within and across specific media platforms and how those features will facilitate or hinder communication across specific groups, topics, and communication goals.

This chapter provides an overview of what we know, based on a growing body of science communication research in social media settings, about practicing successful engagement with scientific information across a variety of communication goals. The first section begins with a description of why and how scientists and science communicators use social media: The features of social media that draw communicators to engage with these platforms, including the incentives that scientists and science communicators themselves perceive for using social media in their work or leisure time, and how communicators frequently adapt their approaches across different social media platforms. The chapter then moves in the second section to the pros and cons, or particular features to be aware of, among different platforms. Given the high turnover rate of social media platforms in general and of different platforms' specific features, this section paints a more broad-brushed illustration of aspects of platform design to consider when communicating through any platform, using examples from current prominent or growing platforms such as Facebook, Twitter, YouTube, Instagram, and reddit. Finally, the third section provides an overview of how science communication research and training help inform best practices on social media, and how communicators can benefit from and contribute to these areas to strengthen future science communication.

Why and how science communicators engage through social media

Beyond the reinforcing interactions of both interested publics and science communicators increasingly moving to online media to communicate and find information, several important features of social media incentivize using these platforms for communication. First, it is important to note that literature on communication and engagement often makes a useful distinction between one-way versus two-way communication, with the latter describing a higher level of engagement, as well as some typologies distinguishing between "communication" and "engagement" (see Irwin, 2014; Rowe & Frewer, 2016 for overviews). In practice, the important components that define communication and engagement – providing information and interacting with and listening to others – are often mixed together in different amounts, especially in social media settings. It is not the purpose of this chapter to explicitly define communication versus engagement. Instead, we will use the term "communication" throughout as an all-encompassing term for sharing information, and refer to engagement when highlighting different levels in terms of how frequently people communicate and whether they participate in two-way communication. This component of the level of engagement a communicator wants to achieve or participate in is important to understand when picking communication goals and strategies, and is a focus of this chapter.

As mentioned in the introduction, a large incentive for communicating through social media is that these platforms are increasingly where communication space exists, especially around science issues. Many interested publics search for science news online and use social media for news and leisure (Newman et al., 2017; Funk, Gottfried, & Mitchell, 2017), and anyone can create content on social media with relatively little time and resources compared to traditional communication approaches such as face-to-face events, printed communication, and journalistic or documentary TV or radio segments. Not having to rely on more traditional media sources to create and distribute content also means that science communication on social media can incorporate more personal and less formal formats and content (Brumfiel, 2009). Many scientists and science communicators state intrinsic motivations for sharing content, such as the potential for self-expression (Rainie & Wellman, 2012), curiosity and interest in science information (Collins, Shiffman, & Rock, 2016; Ranger & Bultitude, 2016;), and wanting to build peer and public collaborations and communities (Collins et al., 2016). Many regular science information-seekers attend to science-specific blogs and other social media content because of similar motivations (Jarreau & Porter, 2017).

Extrinsic rewards and opportunities to reach those outside of the typically science-interested publics motivate many science communicators to use social media platforms as well. For many communicators, communicating through social media can be part of their career-building strategies. Journalists of legacy and online media often use social media platforms, such as Twitter, for story leads (Broersma & Graham, 2012; Brumfiel, 2009), which can increase the reach of a communicator's work, and social media also has the potential to increase citation rates and attention for scientists' research (Haustein et al., 2014; Jia, Wang, Miao, & Zhu, 2017; Liang et al., 2014; Rainie & Wellman, 2012; Thelwall, Haustein, Lariviere, & Sugimoto, 2013).

Because of the structure of social and information networks online, communication can target a particular group based on interests and characteristics or can spread across multiple diverse networks. As described in more detail in the third section of this chapter on science communication training, this flexibility and varying level of message control can provide incentives for communicators, as well (Jia et al., 2017; Pavlov et al., 2017). The range of types of platforms and communication methods also allows for varying levels of potential engagement with different publics and opportunities to communicate with people who are not the typical science enthusiast. The opportunities for two-way engagement with interested publics (Mahrt & Puschmann, 2014; Yeo & Brossard, 2017) and for making science and aspects of the daily workings of scientific research more accessible and open (Stilgoe, Lock, & Wilson, 2014; Van Noorden, 2014; Yeo & Brossard, 2017; Yeo et al., 2017) and more relevant to public and decision-makers (Allgaier et al., 2013), are strong motivations for many scientists and science communicators who share information through social media (Howell et al., 2018).

Mixed media approach: the pro and cons of different platform features

Although the features of social media platforms can lend themselves well to two-way communication, any particular feature on social media typically has the potential to be both a tool for and a barrier to effective communication. In this section, we outline some of the main features that are especially salient to communication on social media across a wide range of platforms and that research has found can help and hinder communication goals. These, often interrelated, features are: platform and search engine algorithms; technologically and socially created filter bubbles or social fragmentation; user make-up and level of openness/publicness of platform content; social and informational context cues, such as platform-suggested stories, article comments, and post shares; and platform layout and site design.

Algorithms – who sees what, where, and why

Search engine algorithms play a large role in determining what type of content people come across when seeking information on a topic, and social media platforms continually change their algorithms to promote certain content reaching users' front pages (see Oremus, 2016, 2018 for examples of Facebook's algorithm and its effects on content). Part of the effect of algorithms is due to specific company decisions, often designed to meet company goals and maximize profit by providing what the company decides users want to see and are likely to engage with (Scheufele & Nisbet, 2013). Part of the effect, then, is also due to how the algorithms respond to information from users, who, through their choices in what links and posts to engage with, shape what the algorithm decides is popular and relevant (Scheufele & Nisbet, 2013). These two factors intertwine and affect each other, and the result is that particular information becomes more or less accessible to particular users due to this structural component of online, and particularly social, media (Ladwig, Anderson, Brossard, Scheufele, & Shaw, 2010; Li, Anderson, Borssard, & Scheufele, 2014; Liang, Anderson, Scheufele, Brossard, & Xenos, 2012).

Understanding the commercially protected black boxes that shape algorithmic outcomes is beyond the scope of this chapter and of science communicators' work. Communicators should, however, do basic background research on how the specific platforms organize posts (e.g., chronologically, in which case frequent posting matters more, or by popularity or "relevance," in which case using the right key words and tags is especially important) before sharing content on those platforms. These algorithms do matter, though, in one way that is especially relevant for communicators using social media, and that is the potential for fragmentation online.

Online fragmentation – scope and communication effects

Because algorithms online are designed to produce content that users want to see, this content often aligns with users' short-term personal interests and preferences. The result is that algorithm-enforced user choices can create filter bubbles – or fragmentation and isolation based on interests and opinions – in online communication settings (Pariser, 2011; Scheufele & Nisbet, 2013). Algorithms and other structural components of online information environments, such as links between pages, also typically create a "power curve" structure of online content. This power curve is characterized by a few well-linked and heavily trafficked sites at the top of the curve and a long tail of less linked and visited sites (Benkler, 2006; Hindman, 2009; Webster & Ksiazek, 2012). If sites at the tail are not well-connected, that could be evidence of fragmented information environments living in relative isolation from each other.

Much of the concern around fragmentation relates to political discourse and civic deliberation (Garrett, 2009; Newman et al., 2017), but it has implications for science communication as well. Science issues that overlap with political divisions or polarizing policy decisions, such a climate change, could be especially vulnerable to online discourse that exists in filter bubbles (Yeo, Xenos, Brossard, & Scheufele, 2015). On the other hand, science-related content that does not overlap with broader public interests in terms of societal or more human-focused or entertainment-related content could exist only within the information bubbles of a small group of science enthusiasts or interested publics (Xenos, Becker, Anderson, Brossard, & Scheufele, 2011).

Evidence suggests, however, that although filter bubbles can and do exist, there is also substantial cross-pollination between different information sources, even across societal and interest-based divisions. Social media in particular can encourage this cross-cutting exposure. Within the power curve, there is high levels of overlap and communication between sources in the tail and at the peak (Webster & Ksiazek, 2012). Because of the mixed nature of social media, as well, people who use social media, although exposed to filter bubbles in some cases, also experience more diversity in sources than do people who do not use social media (Flaxman, Goel, & Rao, 2016; Newman et al., 2017). This mixed bag of fragmentation and cross-pollination is partly due to the user make-up and "publicness" of particular social media platforms, which this section describes next. It is also due to how social cues on social media platforms play a role in what information algorithms select and what information people pay attention to and share, which this section will describe below.

User make-up and platform "publicness"

Who uses a platform, and what level of in-group versus broader public communication that platform facilitates, plays a key role in how information spreads and on the impact of that communication material. Platforms that depend more on personal social networks, such as Facebook, will of course have a different reach for each communicator and depend heavily on the communicator's personal network. These more personal and private networks, then, can be less suitable for broad engagement (Collins et al., 2016) and better for sharing with friends and very interested publics who follow specific organization pages (Pavlov et al., 2017). Personal social networks shape most social media platforms, but many, such as Twitter and reddit, can also facilitate a wider range of connections that only exists in online settings, including more professional connections and interactions with journalists or other practitioners and publics that one would otherwise not meet in person (Collins et al., 2016; Broersma & Graham, 2012; Guidry, Jin, Orr, Messner, & Meganck, 2017).

It is clear that across all social media platforms science-related content will often likely be accessed by people who are especially interested in a particular topic or in science more broadly (Jarreau & Porter, 2017). Evidence suggests, however, that the interconnections between different social and more traditional media sources and networks can facilitate reaching a more diverse range of people, if this is the goal (Howell et al., 2018). The final section of this chapter will focus more on the level of openness and user make-up across platforms and their importance for picking the right tools depending on communication goals. We mentioned these dimensions briefly here, however, because they matter for the next major feature that helps determine how information moves within and across social media and larger communication environments, namely the social cues and informational context that surround and shape interpretation of a post on social media platforms.

Social and informational context cues – from "nasty effect" to deliberative engagement

Beyond the original content, all social media have user-created content and algorithm and site-design features that influence the portrayal and interpretation of this content. Comments, for example, are a substantial source of additional information that surround a story or post and can significantly affect how people interpret that post. On the damaging side, uncivil comments can negatively affect how people perceive the trustworthiness of the original post and the issue covered, regardless of the actual content of the post itself, a phenomenon called "the nasty effect" (Anderson, Brossard, Scheufele, Xenos, & Ladwig, 2014; Anderson, Yeo, Brossard, Scheufele, & Xenos, 2018). Comment moderation, however, can alleviate this negative effect. Experimental evidence suggests that even just the appearance of moderation – by telling readers that comments are moderated – reduced the effects of present uncivil comments to levels comparable to that of civil comments (Yeo et al., 2019).

Moderating comments or giving cues that comments are sometimes moderated can be an important alternative to banning comments because, on the more positive side, comments offer opportunities for productive communication and engagement, which can increase deliberative outcomes. When a reporter engaged with comments related to their news article on a local news station's Facebook page, for example, comments were less likely to be uncivil and more likely to be deliberative and evidence-based than when there was no engagement (Stroud, Scacco, Muddiman, & Curry, 2015). Additionally, commenting on a story or sharing related information can provide opportunities to effectively correct misperceptions around a science issue (Bode & Vraga, 2015) and to recommend additional sources of information that individuals might be more likely to pay attention to because they are coming from someone they feel they know, trust, or have directly engaged with (Turcotte, York, Irving, Scholl, & Pingree, 2015).

Social cues – such as who shares the information and the frequency of likes, "loves," shares, and comments – are strong endorsements for people to pay attention to. These cues can provide a proxy for how relevant someone perceives the story to be and are often stronger for motivating people to attend to particular information than are other identity-based cues, such as political ideology (Messing & Westwood, 2014). Understanding how a particular social media platform creates opportunities for these types of supplemental information and cues around a post, therefore, can improve the effectiveness of communication. These features (sharing capabilities, moderated commenting, etc.) provide points for additional productive communication and can help a specific strategic communication reach more, and more varied, people.

Site layout and post design options

Finally, different social media platforms allow for different levels of creativity in post design and have different site layout features that can have impact (the final section will focus in more detail on these features). In general, the mere fact that science content is provided and available online can increase learning (Su, Cacciatore, Scheufele, Brossard, & Xenos, 2014) and even help close knowledge gaps, by lessening the effects of inequity in opportunities for science learning between people of higher and lower socioeconomic status (Cacciatore, Scheufele, & Corley, 2014). What platforms communicators use, and what type of content they produce, should then be based on deliberate choices according to the communication goals: who the communicators want to reach, what information they want to share, and why they are communicating, or what outcomes they want to achieve. These considerations are the focus of the next section, which examines applying communication research to affective science communication training and practice.

Science communication training: a bridge from research to practice

Communicating science on social media is similar to communicating in any environment in that communicators have to be deliberate in their choices and, when possible, test and update approaches to meet communication goals. In this section, we will provide brief recommendations and points to consider for developing effective communication strategies on social media platforms, based on empirical research findings. (Note that the National Academies of Sciences, Engineering, and Medicine's report on *Communicating Science Effectively* (National Academies of Sciences, 2017) provides useful guidelines for science communication researchers, trainers, and practitioners on for thinking about in-person versus mediated communication settings.)

Because of the variety of social media platforms and their particular features, as described in the second section of this chapter, science communication will be most effective when using a multi-pronged approach customized for particular purposes and mediums (Bik & Goldstein (2013) provide a helpful chart for choosing an existing media platform depending on a communicator's motivation for communication and desired level of time commitment). Because specific platforms and their features change over time, however, often on short timescales, it is important to focus less on a particular platform and more on its particular characteristics at the time of communication. These include: User make-up on different platforms at that point in time; how public or private (or controllable) the sharing opportunities are on a platform; what formats one can use to share information (e.g., pictures, long-form stories, short updates); how the site typically sorts information both on the front page and in terms of directing search traffic; what comments and related information and social cues make up the information context; and platform layout design.

Here we sort these into four broader areas to consider when designing communication strategies:

1 the communication goals – with whom and for what purposes;
2 the level of engagement and resources for engagement;
3 the topic of interest and type of information format;
4 the platform layout and design, including two-way engagement features.

Communication goals – with whom and for what purposes

As with any communication approach, the first step is making explicit what the goals of the communication are – in the sense of how it meets broader intrinsic personal or societal motivations – and what short- and long-term success outcomes will look like. Different goals will involve reaching different people. As described in the second section, reaching different people will depend on what platform one chooses and how one crafts the message. Blogs, for example, are traditionally better at reaching those who are already interested in and seeking out a topic (Jarreau & Porter, 2017). Most other platforms, however, have higher potential for facilitating incidental exposure to science information (e.g., the front page of reddit) or cross-cutting diverse publics (e.g., highly socially endorsed and broadly personally relevant information on Facebook) or being browsed by mainstream journalists (e.g., Twitter).

When it comes to crafting the message, the language communicators use and the angle or connections they include will shape how accessible and relevant the messages appear to different people. As mentioned above, expanding the topic beyond "science" into more broad human aspects, social issues, or topics in which many people have a common interest, can be excited about, or can see as relevant to their everyday lives, can be more effective for facilitating more public discourse and gaining greater attention (Southwell, 2017; Xenos et al., 2011). If the topic is potentially controversial among the groups that one wants to communicate with, communication missteps can have the opposite effect of what the original communication goal was. It is therefore especially vital that this step of crafting the message involves pre-testing and engaging with the communication and public opinion literature on the topic. Fortunately, controversial issues, such as climate change, vaccinations, and genetically modified organisms, also tend to be the ones for which we have the greatest amount of research on communication approaches and effects, including through social media, to inform communication practice.

Level of engagement resources: time, money, and experience

Depending on the communicator's level of interest in two-way engagement and amount of time and resources available to dedicate to engaging, different platforms and types of posts will be a better match. Planning should involve deciding how frequently one wants to or can post and how much energy to dedicate to moderating and engaging in comments and discussion around the post. These decisions also relate to how publicly accessible communicators want their post to be. Posts that are more public likely will require more work. This is because of both the initial preparation time needed and the subsequent engagement required for effectively communicating in public settings. Initial preparation could be more involved because the message is intended to reach a broader group and therefore requires careful crafting of a message to be effective for the range of potential people who could see or engage with it. Moderating and helping keep the message on point as it spreads through subsequent discourse could then also require more immediate and greater effort in facilitating discussion, responding to comments, and addressing requests for more information or countering misinformation.

Communicators who are concerned about not having the time or resources to effectively communicate or who want to be part of a larger communication effort can work with communication teams at their workplaces or research institutions. These teams can have expertise in initial communication strategy and messaging, continued engagement, and trouble-shooting (Chan et al., 2017). If a communicator does not have institutional support, they can still find ways to engage in the desired way given the resources and time they have. Platforms such as Twitter, for example, might be more effective if one can post and respond frequently (Pavlov et al., 2017) but require less time for each single post. Communicators can also reach out to people who run science engagement blogs or YouTube channels or moderators on sites such as reddit's Ask Me Anything thread to create collaborations for one time or more frequent engagement.

Topic of interest: controversial, obscure, photogenic, or abstract

Moving to the format of the message itself, different topics lend themselves well to different types of content. This decision overlaps, as well, with whom communicators want to communicate with. For example, if communicating an obscure or abstract science topic or concept to a broader range of publics, it can help to connect it to more familiar concepts or shared values and priorities. Finding space for long-form communication for interested publics or combining a short and simple Twitter message to achieve more general and broad communication with a link to more information for those who are interested could be other useful approaches. If a topic is particularly photogenic, such as areas of astronomy and ecology, using resources like Instagram or video formats, could be an effective way to reach both more and less scientifically-engaged publics (Pavlov et al., 2017). Communication around a topic that is current and being frequently discussed or updated with new information can work well on faster-paced and often chronological formats, such as reddit posts and comments or Twitter.

Platform layout and social and informational cues

Finally, additional informational and social cues significantly affect information intake and interpretation and offer potential points for additional communication and engagement. These include comments, sharing related stories from trusted or known sources, algorithm-generated related stories that follow a post, and social endorsements in the form of likes and shares. Depending on time and resources for monitoring and contributing to this engagement, different platforms will be more fitting for different goals and time commitments. Twitter, for example, operates more by frequency and currency of posts, which could require more willingness to continue to engage as a discussion runs its course, and sometimes is rediscovered and comes back to life. Each engagement, or Tweet, however, would likely require less time than creating a longer piece of content, such as for blogs or YouTube. Communicators should become familiar with the different platforms and spend some time as a bystander or with smaller engagement commitments to figure out the features and types of conversations that they will want to account for when planning communication.

During this scouting period, it is also useful to pay attention to the particular page design of these platforms. Layout of information in media affects how and what people learn from the content, which matters for communication goals and outcomes. Linear design can facilitate factual learning, while nonlinear design can increase how much a reader perceives the involved issues, actors, and concepts to be interrelated (Eveland, Cortese, Park, & Dunwoody, 2004). Both could be desirable goals, and one might be more appropriate for particular topics or desired outcomes than the other. For example, if one is trying to highlight how a particular science issue relates to a broader picture, a nonlinear design might be more conducive.

Additionally, websites that show one story at a time, such as a blog page, can encourage more directed learning from a particular article but less time reading and exposure to fewer articles (Kruikemeier, Lecheler, & Boyer, 2017). A newspaper format, on the other hand, with multiple stories next to each other (which arguably is what many social media sites are more similar to) can increase the amount of time people spend reading and increase recall of the information across a more diverse set of topics (Kruikemeier et al., 2017).

We need more research into these areas, but communicators can conduct miniature official or unofficial data collections themselves by keeping track of social engagement information (such as likes and shares) across different types of layouts and content styles. Communicators who regularly contribute content through a YouTube account, for example, or an organization's Facebook page can contact the platform companies to see what information companies provide access to, to help in evaluating how much reach a communication piece has on a given platform. Doing this evaluative work, as much as possible given one's time and resources, is necessary for understanding if a particular communication campaign or strategy is effective. What works in one context might be a waste of time (unless a researcher just loves writing and sharing for intrinsic purposes) or actually worsen communication efforts in another.

This evaluation work is where collaborations with science communication researchers can be especially fruitful for both communicators and researchers. Each group has resources and experience that can complement those of other groups. Such collaborations can advance rigorous communication research, ensure more successful communication practices, and create stronger science communication training. Given the fast rate at which social media has grown, building on the important science communication research so far and creating new partnerships will be vital for creating effective communication in the increasingly changing and important social media environment.

References

Allgaier, J., Dunwoody, S., Brossard, D., Lo, Y.-Y., & Peters, H. P. (2013). Journalism and social media as means of observing the contexts of science. *Bioscience, 63*, 284–287.

Anderson, A. A., Brossard, D., Scheufele, D. A., Xenos, M. A., & Ladwig, P. (2014). The "Nasty Effect:" Online incivility and risk perceptions of emerging technologies. *Journal of Computer-Mediated Communication, 19*, 373–387.

Anderson, A. A., Yeo, S. K., Brossard, D., Scheufele, D. A., & Xenos, M. A. (2018). Toxic talk: How online incivility can undermine perceptions of media. *International Journal of Public Opinion Research, 30*, 156–168.

Benkler, Y. (2006). *The Wealth of Networks: How Social Production Transforms Markets and Freedom*. New Haven, CT: Yale University Press.

Bik, H. M., & Goldstein, M. C. (2013). An introduction to social media for scientists. *PLoS Biology, 11*, e1001535.

Bode, L., & Vraga, E. K. (2015). In related news, that was wrong: The correction of misinformation through related stories functionality in social media. *Journal of Communication, 65*, 619–638.

Broersma, M., & Graham, T. (2012). Social media as beat. *Journalism Practice, 6*, 403–419.

Brossard, D. (2013). New media landscapes and the science information consumer. *Proceedings of the National Academies of Science, 110*, 14096–14101.

Brossard, D., & Scheufele, D. A. (2013). Science, new media, and the public. *Science, 339*, 40–41.

Brumfiel, G. (2009). Supplanting the old media? *Nature, 458*, 274–277.

Cacciatore, M. A., Scheufele, D. A., & Corley, E. A. (2014). Another (methodological) look at knowledge gaps and the internet's potential for closing them. *Public Understanding of Science, 23*, 376–394.

Chan, T. M., Stukus, D., Leppink, J., Duque, L., Bigham, B. L., Mehta, N., & Thoma, B. (2017). Social media and the 21st-century scholar: How you can harness social media to amplify your career. *Journal of the American College of Radiology, 15*, 142–148.

Collins, K., Shiffman, D., & Rock, J. (2016). How are scientists using social media in the workplace? *PLoS One, 11*, e0162680.

Eveland, W. P., Cortese, J., Park, H., & Dunwoody, S. (2004). How web site organization influences free recall, factual knowledge, and knowledge structure density. *Human Communication Research, 30*, 208–233.

Flaxman, S., Goel, S., & Rao, J. M. (2016). Filter Bubbles, Echo Chambers, and Online News Consumption. *Public Opinion Quarterly, 80*, 298–320.

Funk, C., Gottfried, J., & Mitchell, A. (2017). Science news and information today. Pew Research Center.

Garrett, R. K. (2009). Echo chambers online?: Politically motivated selective exposure among Internet news users. *Journal of Computer-Mediated Communication, 14*, 265–285.

Greenwood, S., Perrin, A., & Duggan, M. (2016). Social media update 2016. Pew Research Center.

Guidry, J. P. D., Jin, Y., Orr, C. A., Messner, M., & Meganck, S. (2017). Ebola on Instagram and Twitter: How health organizations address the health crisis in their social media engagement. *Public Relations Review, 43*, 477–486.

Haustein, S., Peters, I., Sugimoto, C. R., Thelwall, M., & Larivière, V. (2014). Tweeting biomedicine: An analysis of tweets and citations in the biomedical literature. *Journal of the Association for Information Science and Technology, 65*, 656–669.

Hindman, M. (2009). *The Myth of Digital Democracy*. Princeton, NJ: Princeton University Press.

Howell, E. L., Nepper, J., Brossard, D., Xenos, M. A., & Scheufele, D. A. (2019). Engagement present and future: Graduate student and faculty perceptions of social media and the role of the public in science engagement. *PLoS ONE. 14*(5), e021674.

Howell, E. L., Wirz, C. D., Brossard, D., Jamieson, K. H., Scheufele, D. A., Winneg, K. M., & Xenos, M. A. (2018). National Academy of Sciences report on genetically engineered crops influences public discourse. *Politics and the Life Sciences. 37*(2), 250–261.

Irwin, A. (2014). Risk, science and public communication: Third-order thinking about scientific culture. In M. Bucchi & B. Trench, (Eds.), *The Routledge Handbook of Public Communication of Science and Technology* (2nd edition). New York, NY: Routledge.

Jarreau, P. B., & Porter, L. (2017). Science in the social media age. *Journalism & Mass Communication Quarterly*.

Jia, H., Wang, D., Miao, W., & Zhu, H. (2017). Encountered but not engaged: Examining the use of social media for science communication by Chinese scientists. *Science Communication, 39*, 646–672.

Kruikemeier, S., Lecheler, S., & Boyer, M. M. (2017). Learning from news on different media platforms: An eye-tracking experiment. *Political Communication, 00*, 1–22.

Ladwig, P., Anderson, A. A., Brossard, D., Scheufele, D. A., & Shaw, B. R. (2010). Narrowing the nano discourse. *Materials Today, 13*, 52–54.

Li, N., Anderson, A. A., Brossard, D., & Scheufele, D. A. (2014). Channeling Science Information Seekers' Attention? A Content Analysis of Top-Ranked vs. Lower-Ranked Sites in Google. *Journal of Computer-Mediated Communication, 19*, 562–575.

Liang, X., Anderson, A. A., Scheufele, D. A., Brossard, D. & Xenos, M. A. (2012). Information snapshots: What Google searches really tell us about emerging technologies. *Nano Today, 7*, 72–75.

Liang, X., Su, L. Y.-F., Yeo, S. K., Scheufele, D. A., Brossard, D., Xenos, M., … Corley, E. A. (2014). Building Buzz: (Scientists) Communicating Science in New Media Environments. *Journalism & Mass Communication Quarterly, 91*, 772–791.

Mahrt, M., & Puschmann, C. (2014). Science blogging: An exploratory study of motives, styles, and audience reactions. *Journal of Science Communication, 13*, 1–17.

Messing, S., & Westwood, S. J. (2014). Selective exposure in the age of social media. *Communication Research, 41*, 1042–1063.

National Academies of Sciences (2017). Communicating Science Effectively: A Research Agenda. Washington, DC.

Newman, N., Fletcher, R., Kalogeropoulos, A., Levy, D. A. L., & Nielsen, R. K. (2017). Reuters institute digital news report. Reuters Institute.

Oremus, W. (2016). Who controls your Facebook feed? *Slate*. www.slate.com: The Slate Group.

Oremus, W. (2018). The great Facebook crash. *Slate.com*, June 27, 2018.

Pariser, E. (2011). *The Filter Bubble: How the New Personalized Web Is Changing What We Read and How We Think*. New York, NY: Penguin.

Pavlov, A. K., Meyer, A., Rösel, A., Cohen, L., King, J., Itkin, P., … Granskog, M. A. (2017). Does your lab use social media? Sharing three years of experience in science communication. *Bulletin of the American Meteorological Society*, June, 1135–1146.

Peters, H. P., Dunwoody, S., Allgaier, J., Lo, Y.-Y. & Brossard, D. (2014). Public communication of science 2.0: Is the communication of science via the "new media" online a genuine transformation or old wine in new bottles? *EMBO Rep*, *15*, 749–753.

Pew Research Center (2016). News use across social media platforms. The Pew Research Center.

Rainie, L., & Wellman, B. (2012). Networked creators. *Networked: The New Social Operating System*. Cambridge, MA: MIT Press.

Ranger, M., & Bultitude, K. (2016). 'The kind of mildly curious sort of science interested person like me': Science bloggers' practices relating to audience recruitment. *Public Understanding of Science*, *25*, 361–378.

Rowe, G., & Frewer, L. J. (2016). A typology of public engagement mechanisms. *Science, Technology, & Human Values*, *30*, 251–290.

Scheufele, D. A., & Nisbet, M. C. (2013). Commentary: Online news and the demise of political disagreement. *Annals of the International Communication Association*, *36*, 45–53.

Southwell, B. G. (2017). Promoting popular understanding of science and health through social networks. In K. H. Jamieson, D. M. Kahan & D. A. Scheufele (Eds.), *The Oxford Handbook of the Science of Science Communication*. Oxford, UK: Oxford University Press.

Stilgoe, J., Lock, S. J., & Wilsdon, J. (2014). Why should we promote public engagement with science? *Public Understanding of Science*, *23*, 4–15.

Stroud, N. J., Scacco, J. M., Muddiman, A., & Curry, A. L. (2015). Changing deliberative norms on news organizations' Facebook sites. *Journal of Computer-Mediated Communication*, *20*, 188–203.

Su, L. Y.-F., Cacciatore, M. A., Scheufele, D. A., Brossard, D., & Xenos, M. A. (2014). Inequalities in scientific understanding. *Science Communication*, 36, 352–378.

Thelwall, M., Haustein, S., Lariviere, V., & Sugimoto, C. R. (2013). Do altmetrics work? Twitter and ten other social web services. *PLoS One*, *8*, e64841.

Turcotte, J., York, C., Irving, J., Scholl, R. M., & Pingree, R. J. (2015). News recommendations from social media opinion leaders: Effects on media trust and information seeking. *Journal of Computer-Mediated Communication*, *20*, 520–535.

Van Noorden, R. (2014). Scientists and the social network. *Nature*, *512*, 126–129.

Webster, J. G., & Ksiazek, T. B. (2012). The dynamics of audience fragmentation: public attention in an age of digital media. *Journal of Communication*, *62*, 39–56.

Xenos, M. A., Becker, A. B., Anderson, A. A., Brossard, D., & Scheufele, D. A. (2011). Stimulating upstream engagement: An experimental study of nanotechnology information-seeking. *Social Science Quarterly*, *92*, 1191–1214.

Yeo, S. K., & Brossard, D. (2017). The (changing) nature of scientist–media interactions: A cross-national analysis. In: K. H. Jamieson, D. M. Kahan, & D. A. Scheufele (Eds.), *The Oxford Handbook of the Science of Science Communication.* New York, NY: Oxford University Press.

Yeo, S. K., Liang, X., Brossard, D., Rose, K. M., Korzekwa, K., Scheufele, D. A. & Xenos, M. A. (2017). The case of #arseniclife: Blogs and Twitter in informal peer review. *Public Underst Sci*, 26, 937–952.

Yeo, S. K., Su, L. Y.-F., Scheufele, D. A., Brossard, D., Xenos, M. A., & Corley, E. A. (2019). The effect of comment moderation on perceived bias in science news. *Information, Communication & Society*, *22*, 129–146.

Yeo, S. K., Xenos, M. A., Brossard, D., & Scheufele, D. A. (2015). Selecting our own science: How communication contexts and individual traits shape information seeking. *The Annals of the American Academy of Political and Social Science*, *658*, 172–191.

Part II

Science communication training design and assessment

5 Training scientists to communicate in a changing world

Toss Gascoigne and Jenni Metcalfe

Introduction

> In an ideal world, communication should be a natural part of the scientific process. But in the practical world we all inhabit, it's clear that some aspects of the communication process take a special effort by scientists and research organizations. The most difficult element…is dealing with the public, and this is often regarded by scientists as an extra task which interferes with the real work of research.
>
> (Gascoigne, 2006)

To cope with this challenge, scientists need to be trained. We conducted our first training workshop in 1992. It was held in Rockhampton, a regional town in northern Australia, for a group of 12 scientists working in the Commonwealth Scientific and Industrial Research Organisation (CSIRO) Division of Tropical Crops and Pastures. The subject was media skills and the workshop ran over two days. We jointly led the workshop, and four journalists came in to explain how they worked and to give the scientists practice in interview technique.

This was the first of about 1700 workshops we have run since then. We have been back to Rockhampton several times, but our reach has extended across Australia and to about 20 countries internationally, from Colombia and Kenya, to Ethiopia, Germany, Papua New Guinea and across the Pacific.

The workshops have changed with experience and feedback, but some principles have never varied: a hollow-square set-up to encourage discussion; highly practical sessions where participants put into practice what they have just been taught; two presenters to give the sessions freshness and vitality as well as provide individual feedback; and footage of their performances being given to all participants. All our workshops focus on the intended audience for communication: what do they know already? What do they want to find out? Are they comfortable with technical language?

The range of workshop topics has expanded.[1] There are workshops in presentation skills, in writing for the reader, planning communication, using social media, managing media crises, and (more recently) using smart devices to record and edit science footage. There are weeklong master classes combining a range of these subjects for international groups.

The equipment has changed. Initially we packed a large video camera into the baggage of planes along with a stack of VHS tapes: Now the camera is much smaller, carry-on luggage, and participants are given a record of their performances on a USB drive. Workshops are one-day rather than two as time pressures on scientists have increased.

How and why the workshops began

In the years leading up to the first workshops, scientists were encouraged to talk about their work in public. Governments needed to justify increased expenditure on science, because lack of public support posed potential political risks. Research organizations and scientists had good reason to explain the benefits of their work: to justify public funding, prepare people for impending changes, and to pitch for further work. The public wanted to be assured that funds were being spent wisely, that the research was in their interests and ethical sensitivities were respected, and to satisfy their curiosity.

Successive national ministers for science urged scientists to be more active in the media. Ministerial advice has sometimes been couched gently:

> I appreciate having at my fingertips the information I need to argue for maintaining CSIRO's funding levels. I must have an active information flow and it's always useful to see well-timed CSIRO stories in the media to add support to our case;[2]

sometimes as blunt advice: "[Scientists] have got to go out and to sell themselves";[3] and sometimes as provocative abuse: "scientists are wimps".[4]

CSIRO, the largest research organization in Australia, similarly encouraged their scientists. In 1990, Dr. John Stocker, CSIRO CEO, said in launching Project Ambassador:

> Communication is the responsibility of every CSIRO member. In support of our upcoming budget negotiations I invite all staff to become involved in an all-out information campaign to demonstrate the value for money Australia derives from its premier research organisation.[5]

This echoed similar statements by the CSIRO CEO 15 years earlier (Eckersley & Woodruffe, 1984).

In part, this was to counter concern about public understanding and attitudes to science (Eckersley, 1984). After reviewing six surveys of popular attitudes to S & T in Australia, Richard Eckersley concluded that:

> Australians applaud technological process and fear it...we generally regard science and technology as a good thing, but feel threatened by their growing and seemingly uncontrolled power ... this anxiety may be heightened by the fact that few of us feel we are very well informed about science and technology
>
> (Eckersley, 1987, p. 1)

Factors outside the media such as education, direct experiences with technological processes or products, and observable links with science influenced these attitudes, although Ian Barns (1989, p. 22) claimed that "mass media representations are probably the most important continuing influence shaping perceptions of science and technology." Dorothy Nelkin (1990) made a similar observation about the US.

If scientists did not communicate, the consequences could be far-reaching. Julian Cribb, science writer for *The Australian* newspaper from 1990 to1996, provided a sobering scenario:

> The price of insufficient attention to the publicizing of science is high. It can starkly be seen in:
>
> - decline in real government funding and support for scientific research through the 1980s–1990s
> - the drift of bright young Australians away from science studies and careers
> - loss of Australian scientific discoveries overseas, because local firms cannot see the opportunities in commercializing them
> - rundown in the scientific infrastructure – buildings, equipment and technical support
> - low technological awareness in the population at large, leading to a low-tech society which exports cheap and crude and imports dear and sophisticated
> - long term decline in national wealth and living standards.[6]

So scientists were encouraged to communicate their work through the media and the media were ready and waiting (Metcalfe & Gascoigne, 1995). But many scientists were baffled by the media. They did not understand how journalists chose the stories, or what stories interested them, or the mechanics of collecting the ingredients of a story, or the savage deadlines journalists work to. They lacked the necessary skills and experience to be effective in the media. Training in media skills was a pathway.

Ministers, CEOs, and commentators from both Australia and overseas agree: Scientists need to communicate their work to the public. It became a matter of helping them, and over 1990 and 1991, CSIRO offered scientists a course to improve their media skills. CSIRO Corporate Public Affairs contracted TV journalist Patrick O'Neill to run one-day courses for its scientists with both practical and theoretical components. These courses were subsidized by Public Affairs, and cost participants $100.

O'Neill left CSIRO early in 1992 to return to full-time television and his departure left a void. In the absence of any training program, a few of the 35 CSIRO divisions went to outside media consultants, and others offered training sessions run by their communication managers. (It is important to recognize that the divisions then had a good deal of autonomy in matters such as media training for their staff.)

Our workshops began in 1992 when Jenni Metcalfe decided to provide media skills training for scientists in her CSIRO Division, Tropical Crops and Pastures. She wanted a workshop especially tailored for scientists, with some appreciation of the environment in which they worked. Commercial options were not suitable, so she approached Toss Gascoigne, a colleague at CSIRO's Centre for Environmental Mechanics in Canberra, with the suggestion they jointly develop a course.

We developed in our media skills workshops a format and style since applied to other workshops in areas including presentation skills.

Media skills workshops for scientists

The aims of the workshop

We had some experience with the media. Jenni is a university-trained journalist and had worked with various media outlets before joining CSIRO. Toss had previously worked as an occasional freelance journalist. At CSIRO we were involved in launching new programs such as "Clever Clover," the agricultural pasture seed-sowing machine, the bandseeder, and the Coastal Zone Program; and we both wrote media releases and worked closely with journalists.

The aims of the workshop were simple. Our first was to lift the veil of mystery from the media. Scientists who have little contact with media operations do not know how it works. They have no real understanding of the importance of pictures to a television news story; or the rigid formula which applies to most stories in the media; or that the immediacy of a radio story is such that 20 minutes later may be too late.

Our second aim was to introduce scientists to real working journalists. Here were two groups with strong stereotyped images: When scientists thought of themselves in the media, they pictured being grilled by a hard-hitting shock-jock; when journalists thought of scientists they summoned images from *Revenge of the Nerds*. It added up to a relationship of mutual suspicion. We wanted to bring scientists and journalists together, act as an introduction agency, and show that despite the differences there could be a mutuality of interest.

The third aim was to give the scientists practical experience of being interviewed. Practice and a little knowledge in print, radio and television can make a powerful difference. A scientist armed with five tips on how to handle a television interview and 10 minutes' experience from a training session can approach a real interview with much more confidence.

Constructing the program

We drew on many sources in constructing the program. One function of our jobs was to act as facilitators and intermediaries for scientists to get their stories into the media, and that experience helped set the broad agenda. Our previous journalistic experience was useful, as were related courses.

We looked at relevant literature: *Communicating Science* by Michael Shortland and Jane Gregory and *How to Use the Media in Australia* by Iola Mathews were two books we used. To structure the workshops, we drew on our former experience as secondary school teachers, aiming for sessions that were varied and stimulating.

It took several months of research (alongside our normal jobs) to compile notes and create the framework. Content covered media releases and staging a launch, with sections emphasizing the priorities of radio, television and the print media. The notes included copies of media releases, hot tips, and a paper by Julian Cribb (quoted above).

The core of the media workshops is the involvement of working journalists. We quickly learned that scientists really enjoyed meeting journalists and working out how to explain their work to people who generally did not have much science education (most journalists in Australia have backgrounds in the humanities or social sciences). Our practice since the beginning has been to hire working journalists from television, print and radio. They are paid to lead a session where they describe their work and interview the participants.

We organize journalists by ringing up the local stations or newspapers, to see who is available. We like a mixture of journalists, science specialists and general journalists who might cover any story: an election, an earthquake disaster, a murder – or a science story. It is important for scientists to learn to tell their story to both specialist and generalist journalists. One benefit of employing working journalists is that it keeps the workshops up to date. The media changes constantly and although as presenters we have a good background knowledge, we are not always familiar with the latest developments.

The workshop format

Media skills workshops now normally run for one day rather than two, 9 a.m. to 5 p.m., with a normal maximum of 10 participants. They emphasize the practical: All participants will be interviewed at least three times on their story. They are also informal: We encourage the scientists to ask questions and to bring up issues. It is not a series of lectures but discussions and practical exercises.

The participants sit in a hollow square. We bring a video camera and a tripod and there is a screen and a whiteboard at the front of the room. All participants are given a 60-page booklet of notes. A week before the workshop, they send us a brief outline of the story they want to work on during the workshop.

The workshop program

We begin by asking participants which issues are most important, to choose their top three from a list of a dozen items, including:

- understanding the pressures and constraints under which journalists operate;
- tailoring a scientific message to suit the media, without compromising the quality of the science;

- gaining experience in media interviews (TV, radio and print);
- dealing with difficult questions;
- finding out what journalists need to make a story work.

Almost invariably scientists choose the second option. The first option about journalists is almost never selected, although post-workshop feedback often nominates interactions with journalists as the highlight. We tweak the program to meet the priority issues participants have chosen.

The first journalist arrives after a preliminary discussion, about an hour into the workshop, and the session begins with a conversation on TV news:

- How many stories do you do each day?
- How do you choose the stories? Where do the ideas come from?
- How long does the interview with a scientist normally last?
- What questions do you ask?
- How do you get all the pictures and footage you need?
- How long is a typical story?

After 30 minutes the journalist, the camera and one of the presenters moves to a second room, and participants go out one by one for a short (5-minute) interview, with feedback from the journalist.

While the interviews are going on, the remaining participants prepare a simple description of the most important or interesting aspect of their story. At 11 a.m. the second journalist arrives, from the print media. Each participant will then read out their story idea, and the journalist asks questions about the research and its implications, before giving feedback on where the story might fit into the newspaper: the news pages, or a specialist section or a feature story.

When the individual television interviews are completed and participants are all back in the main room, the print journalist leads a discussion:

- How is print different to television?
- How important are photographs?
- Do you do interviews over the phone or visit the scientist?
- What does a sub-editor do, and do they ask you before changing your original story?
- Who writes the headlines?
- How do you choose which stories to write, out of all the emails and tweets and media releases that come in?

The print journalist is more likely to be a science specialist than those from TV or radio. They may not have formal science qualifications, but will have a good working knowledge of science issues and people working in this field.

The third journalist is from radio, and this session follows the same pattern: a discussion followed by individual interviews. While interviews are going on in a separate room, the remaining participants will be engaged in a variety of exercises:

- Working out the main points they want to make in interviews;
- Learning how to stick to the point and not be distracted;
- How to handle questions they can't (or don't want) to answer;
- Learning how the organization's science communicator can help them;
- Handling an ambush interview;
- The value of media releases.

One novel exercise tests participants' understanding of the structure of a television story. They are asked three questions:

1 How long does a typical TV news story run?
2 How long are the "grabs" or "sound bites" of the scientist?
3 How many different pictures does the story use: In other words, how many times does the camera cut from one shot to another?

Their answers are recorded on the whiteboard (and can vary by an order of magnitude). Participants are then shown three science stories from TV news. One participant times the length of the story, a second times the sound bites from the scientist, and everyone else counts the number of shots. The question almost everyone gets wrong is the number of shots with participants invariably underestimating how many pictures make up 90 seconds of TV news.

The purpose of the exercise is two-fold: To show that pictures are essential to making a TV story and you need a lot of them; and that TV news is formulaic. All stories follow a format: they are about the same length (90 seconds), the "grabs" from the scientist are between 3 to 12 seconds; and journalists need 25 or so different pictures to make the story. To scientists, this was a surprise:

> The one lesson that did take me by surprise was the blindingly obvious point that television and print media are interested in pictures. Watching several news items and counting the number of different shots … soon convinced us that visuals are the story when it comes to TV.[7]

But scientists are logical people and once they understand the structures, they can exploit them to great effect.

What scientists find difficult to do

Scientists have a special difficulty with the media. In many ways, their training and experience is at odds with the media. For instance, years of education have drummed into scientists that they should answer questions directly, fully and accurately. But their own best interests during an interview – particularly with electronic media – might be served by circumventing the question in favor of repeating the main point they want to get across. They may go into too much detail and risk confusing the journalist. Or they can be too honest for their own good, and hesitate fatally in front of the microphone or camera when asked career-threatening questions like: "Do you think the government is doing enough in this area?"

Our participants are startled when they are told *they should know what they want to say before a media interview starts*. Their normal interview technique is to sit back and wait for the questions, thinking of it as an intelligent conversation and forgetting that *they* are the experts and the journalist probably knows very little. We point out that it is the scientist's responsibility to make sure the story comes out the way they want, and they need to be on the front foot. This may require some gentle deflection: "You have raised an interesting question there, but the main point about my work is that...". We are not teaching them to become like politicians; rather they need to learn that if they want to get their prepared story and message out they need to have a measure of control.

Scientists and journalists come from different worlds, summed up neatly by Michael Shortland and Jane Gregory (1991):

> Scientists see science as a cumulative, co-operative enterprise; journalists like to write about individual scientists who have made a revolutionary breakthrough. Journalists like controversy; scientists thrive on consensus. Journalists like new, even tentative results with exciting potential; scientists prefer their results to go through the slow process of peer review and settle into a quiet, moderate niche in the scientific literature – by which time journalists are no longer interested. Scientists think that accuracy means giving one authoritative account; journalists feel that differing views add up to a more complete picture. Journalists' work has to fit the space available; scientists' academic papers can be any length. Scientists work at the pace imposed by the nature of the research; journalists are in a hurry to meet a deadline. Scientists must qualify and reference their work; journalists have to get to the point.
>
> A one-day workshop can be transformative.
>
> (Shortland & Gregory, 1991, p. 144)

Participants learn the media is interested in how their work will affect the audience watching and listening – more jobs, a cheaper loaf of bread, better environment – but not in how they performed the research. They hear television journalists say any potential story is viewed through the lens of "Betty from Blacktown," a mythical 50-year-old woman from a working-class suburb and typical of the television audience. If journalists think the story would interest Betty, they will cover the story.

The media skills workshop we have developed is practical, up-to-date and tailored for scientists and their culture. The workshops have helped many people, but there are other incentives and disincentives scientists face in using the media to communicate with the public.[8]

Assessments and evaluations

At the end of each workshop, we ask participants to complete an evaluation form. The first six questions ask them to rate aspects of the workshop on a 1 to 7 scale, where seven is high:

- 1: Overall assessment of the course;
- 2: Course content, information and ideas presented;
- 3: The presentation/facilitation style of the consultant;
- 4: Mix of information, presentation, discussion and activity;
- 5: The likely usefulness of the workshop notes in the future;
- 6: Recommendation of this course to others at a similar level.

The next three are open-ended: "What did you like most about the workshop? Is there anything you would change? Any other comments?" Participants can opt to be anonymous. Feedback is almost invariably positive with average numerical scores normally ranging between 5.8 and 6.5 on a 7-point scale (any score below 5 makes us stop and consider what went wrong). Comments are also normally positive:

> It was interesting to get insights into journalists, their job, their pressures, what sells a story and how best to do it. Being able to talk to working journalists and see them as people not to be feared was the highlight.
>
> (Anonymous comment on a feedback sheet, 1998)

We send an unedited compilation of comments to the organization commissioning the course, and use the feedback to refresh, revise and modify our approach. Changes over the years include: shortening and simplifying workshop content, hiring more journalists for media workshops, more emphasis on the practical and less reference to the workshop notes, a strict policy of finishing on time, incorporating variety into the program (exercises, discussions, individual work), and careful room preparation to ensure the most comfortable possible space.

At the end of a workshop we ask which form of media participants prefer. Responses are almost equally divided: Some prefer print, seeing it as more detailed and more "permanent," others prefer television because of its impact, while the third group likes radio because of its immediacy and simplicity.

A second form of evaluation tests how meeting journalists affected the attitude of scientists. Were their views more positive after talking to journalists and being interviewed by them? For a series of workshops over the period of a year, we distributed a sheet with 12 words at the beginning of the workshops, and asked participants to rate journalists on a 1 ("strongly agree") to 7 ("strongly disagree") score for each word. At the end of the workshop, after they had had intensive dealings with five different journalists [these were two-day workshops], they were given an identical set of words and asked to score them again. The sheet contained both positive and negative words:

- Helpful
- Reliable
- Sensationalize
- Trivialize
- Thorough
- Accurate
- Distort
- Superficial
- Interested
- Concerned
- Unprincipled
- Trustworthy

The "after" answers showed participants were much more positive about journalists after meeting them. They were more likely to think of journalists as "helpful," "thorough," "concerned," "reliable," "accurate," "trustworthy," and "interested." The average change measured was a swing of about one in a positive direction. In other words, if a participant had scored journalists as "3" on the word "accurate" at the beginning of the course, at the end they would re-score them as "4."

Some differences were quite dramatic. Nearly 90% of the participants changed their score on the word "sensationalize," all in a positive direction. Seventy percent changed their score on "unprincipled," all positive. The words "distort," "trivialize" and "superficial" also drew strong results (but each with a handful of negative responses) (Metcalfe & Gascoigne, 2001).

The survey showed that meeting journalists improved the attitude of scientists. Did it have the same impact on journalists who led a session at the workshops? We tested these views by asking journalists to complete a post-workshop survey. They too were positive about the experience and the value of the workshops. They were relieved to discover scientists were a potential source of good stories; prepared (to some extent) to make the compromises the media demands; and that talking to scientists broadened their own outlook. As one journalist stated:

> Most of the scientists in the workshop in which I participated had never had much media contact, and they were anxious about dealing with the media. I'm sure we managed to show that really, we're quite nice people, and all we want to achieve is to be able to have a clear and concise chat about new scientific breakthroughs. Easy!
>
> (Metcalfe & Gascoigne, 2001)

In 2017, we asked past workshop participants to complete a survey on the value of the workshop they had attended. Most of the 66 respondents (86%) had attended a workshop in the period 2014–2017; one had attended a workshop in 1996. We asked them to rate the usefulness of the workshop now that some time had passed. More than 80% said it had been highly or very highly useful for communicating with non-technical audiences. Almost 60% also said it had been highly or very highly useful for communicating with their peers, while 75% said they had changed their attitudes to communication. Most (85%) had applied skills gained to at least a moderate extent. When asked why they had changed their attitudes, respondents said:

> It made me evaluate how a non-scientific audience may view scientists, and made me realise that communicating science is something that needs to be made "human." Jargon doesn't work for a non-scientific audience.
>
> The workshop really drove home for me that I need to think about my target audience and what is the purpose of my communication. Only then, can I start to think about how best to deliver the message.
>
> (Anonymous comment in survey response, 2017)

Expansion of skills workshops into other communication topics

The initial media skills workshops in 1992 were well-received and Lindsay Bevege, Director of CSIRO's Public Affairs section, asked us to develop another workshop, focusing on presentation skills and building the ability of scientists to deliver clear and logical accounts of their work to different audiences. Scientists regularly describe their work at seminars and conferences, but they might also expect to talk to business and industry, policy-makers, funding bodies, community groups and school children. The content, tone, style, complexity and the focus have to be targeted to each audience.

The demand for both media skills and presentation skills workshops increased and we were hired out from our CSIRO divisions to run them. CSIRO staff paid a small fee to attend the workshops, enough to cover costs.

The new presentation skills workshops were also a success. They ran on similar principles to the media skills workshops: a maximum of about 10 participants, highly practical in nature, hollow-square room set-up and use of video camera to record and playback performances by participants.

When scientists stand in front of a live audience, they may focus on their own concerns rather than those of the audience. In presentation skills workshops they learn to focus on the audience. What does the audience already know? What do they want to hear? Why have they chosen to attend the talk? Once scientists have worked out the answers to these questions, then they can construct a suitable talk. We suggest they try to structure their story around three main points – "there are three things I want to discuss today" – and resist the temptation to load up on details. Audiences cannot absorb detail in a verbal presentation, there is rarely time, and if they want detail, audiences can be referred to the associated papers. Scientists giving presentations face an ongoing battle between detail, clarity and time. Unfortunately, detail often wins out at the expense of time and/or clarity.

When asked to nominate their priorities for the day from a list of a dozen options, presentation skills participants often choose "How to structure a talk" or "How to handle difficult questions." We ask them back: "What is a difficult question?" and they nominate something off-topic, something outside their area of expertise, or which they can't answer. This is a rare occasion where we suggest a formulaic answer:

> I'm sorry, I can't answer that question [be honest!], because [insert reason e.g. because it's outside the scope of this project, or the results aren't in yet], but what I can say is [choose one of the key points you want to make, as long as it is relevant].

The greatest difficulty scientists face in structuring presentations is writing the introduction and the conclusion. We suggest they begin by saying something challenging to arouse the interest of the audience, perhaps a statement of the problem, and to conclude by discussing the implications of their work. What does it all mean? To do this successfully they have to think in terms of audience: why are people there? What do they want to learn?

In 1995, our employment with CSIRO ended. We both left the organization to pursue other opportunities (Jenni Metcalfe to join a consultancy firm specializing in science and environmental work; Toss Gascoigne to run a national advocacy body for science). Jenni Metcalfe's company, Econnect Communication, took on the workshops on a full commercial basis. What had been in-house workshops for CSIRO were now open to researchers from other organizations, including the Antarctic Division, Geosciences Australia, the Australian Institute of Marine Sciences, and researchers involved in a new (and still developing) research grouping called Cooperative Research Centers. Government departments also commissioned workshops and the demand increased.

We offered new courses. One was an advanced workshop for senior managers who might be expected to face the media under difficult circumstances: politically sensitive science, staff sackings, accidents or deaths at work, failed experiments. Typically, this was run for a small group over four hours, with journalists asked to apply pressure on sensitive areas. We offered an additional service to build on the value of media skills workshops: preparing a draft media release for each participant, together with an assessment of how and when the story might be released.

Client organizations have different requirements from a workshop, so we were (and still are) flexible in constructing a program to meet their needs: perhaps a one-day workshop combining media skills and presentation skills.

Workshops designed for scientists worked equally for other groups communicating technical topics: researchers in the humanities and social sciences, patent lawyers, farmers and consultants in the cotton industry, a medical college, providers of alternative medical therapies, researchers in defense industries and operational staff in electricity generation.

Our writing courses emphasized the importance of plain speaking. Participants are encouraged to be simple and direct in both writing and speaking. Scientific writing can be difficult to read: often written in the third person, it tends to be stuffed with obscure acronyms, complex language and long sentences. The FOG Index[9] is a useful guide: It tests the readability of a passage of writing, based on the length of the words and sentences. Workshop participants are shown how passages can be pasted into an online tool for an instant readout and they become quite adept at predicting the index once they see passages on screen.

We trained people for the annual Science meets Parliament Day, which brings 150 scientists to Canberra for one-on-one meetings with members of the Federal Parliament. The group gathered in a lecture theater for a training session the day before the meetings, and 10 volunteers came to the front. Their task was to explain their work and its importance, in one minute.

A jury sat on the far side of the stage to provide feedback.

After each scientist spoke, the jury (a journalist, a member of Parliament and a lobbyist) was invited to comment. Was the message clear? Was the language suitable? Did the jury understand the significance of the work? Would they fund it?

Parliamentarians have little expertise in science, are pressed for time, and their working day demands they make difficult decisions on skimpy evidence. As a training session it was quick and brutal – and effective: how to give a brief explanation of complex work and its value in simple terms.

As the workshops expanded and sometimes clashed, we needed additional presenters. There are a dozen people in Australia who have been trained by us to lead a workshop, usually through an apprenticeship approach. There are many others (usually communication staff at research organizations) who have sat in on a workshop, collected a booklet and picked up enough skills and confidence to run aspects of the course with scientists in their organization.

There are also communicators based in countries (South Africa, Papua-New Guinea) where at the end of the day we added a "train-the-trainer" session to talk over how our participants could use their newfound skills to run their own training sessions.

Adapting the workshops to operate internationally

By 2000 the workshops were international. The Royal Society of New Zealand (RSNZ), and the South African Agency for Science and Technology Advancement (SAASTA) invited us to run demonstration media skills workshops in New Zealand and South Africa, three workshops in each country. This was a test of their transferability: Would our courses work in countries with their own cultural and media arrangements?

They did succeed, largely because we researched local needs and cultures and used local experts to lead sessions including local journalists. This was crucial: While the broad approach to journalism is the same, the formats, resources and programming arrangements in South Africa and New Zealand are different. We needed local knowledge to ensure the validity of the courses, and the questions we asked journalists in the workshops took on a particular importance.

Attendees included observers from RSNZ and SAASTA as well as the normal dozen scientist participants. In time, this led to local versions of our workshops; based on our structure and principles but adapted to match the style and preferences of local presenters. In South Africa, after a return series of demonstration workshops in 2004, we worked with Marina Joubert, a local science communicator, to adapt our courses to meet the needs of South African researchers.

In 2011 the Crawford Fund invited us to run masterclasses in science communication, initially in Thailand and subsequently in India, Kenya, Fiji, and Ethiopia. The Fund is a national support organization for international agricultural research, partnering Australian expertise with local partners. Masterclasses were designed with Cathy Reade, the Crawford Fund's Director of Public Affairs, and combined elements from different workshops we offered:

- Planning Communication, where participants draw up a plan and schedule for all their communication and outreach activities;
- Presentation skills;
- Media skills;
- Designing and writing communication documents such as fact sheets, and media releases;
- Using social media.

The Crawford Fund sets out the purpose and scope:

> In developing countries, as in Australia, there is a growing appreciation in agricultural research institutions of the need to confirm the efficacy of investment in agricultural science by better communicating research outcomes and research related stories of interest to a broad range of non-scientific audiences in funding agencies, other interested stakeholders such as farmers and extension agencies, and to the general public through the media.
>
> For many developing country institutions, this role falls to science staff who may have no training in communication and no additional communication staff to take on the important communication role. It is proposed that the Crawford Fund support a course to boost the capability of developing country research institutions to communicate their work to stakeholders (funding, partner and government agencies) and the general public.[10]

The initial workshop in Chiang Mai drew participants from Thailand (3 from Chiang Mai and 3 Bangkok), Cambodia (2), Laos (2), Vietnam (2), Bangladesh (2), Pakistan (2), India (2) and the Philippines (2).

Three weeks before these workshops, we ask participants about their experience in communication, from giving presentations, working with the media and writing articles for the popular media. The survey asks them to select from a list of 13 items what they would like to get from the workshop, and to nominate the communication tasks and objectives attached to one of their projects. We had to modify our standard approaches in designing the survey, preparing a booklet and creating a program and the associated exercises, to account for variable (and in some cases, quite limited) abilities in English. This means that issues are stated simply and directly:

- Who do you think your project should be communicating with?
- Who is the most important group or individual?
- What are the three top messages you'd like to communicate?

This information helps us develop a program and workshop materials. Our arrangement with Crawford makes us responsible for providing equipment (video cameras etc.), engaging in-country journalists to lead sessions on the media, and facilitating the five-day workshop.

A Crawford workshop commissioned for the Pacific nations posed a specific set of challenges. Our workshops rely on discussion sessions, but Pacific Islanders can be shy, very softly spoken and seek to avoid disagreement and controversy. Verbal and writing skills are highly variable and there are cultural surprises (for instance, every meeting begins and ends with prayers). Our first aim is to win their trust. Good humor is essential, as are exercises to encourage participants to speak out.[11] We have adjusted the pace and content to cope with variable abilities in written and spoken English and introduced role-playing exercises for variety and effective learning.

There is a constant factor we have to contend with: learning to cope with the unexpected. These happen in domestic courses but more frequently overseas. For example:

- Rooms not set up to our specifications or completely unsuitable (we ran a workshop in Papua New Guinea in an open grass-roofed enclosure on the shores of the Bismarck Sea with a mounted machine gun souvenir on the side, because the room provided was small, cramped and stifling).
- Malfunctioning, broken or forgotten equipment (an unreliable tripod at a workshop in remote outback Australia caused the video camera to topple over, smashing its lens. We borrowed a camera from a passing BBC crew to complete the workshop.)
- Journalists who are late, called away at the last minute, give strange advice, or turn out not to speak English after all.

Overall the principles we use to construct courses for Australian participants have proved highly applicable in other countries. We have not changed our general approach of focusing on learning by discussion and a strong practical component. The use of guest speakers and journalists to lead sessions adds a valuable (and necessary) local knowledge component to workshops.

The future: new media and other changes

Over the last 26 years the communication space for scientists has changed dramatically.

Part of this has been driven by a tightening of resources available to science communication: The ranks of support staff have thinned and scientists have had to rely more on their own resources for communication activities, from creating a communication plan, writing media releases and contacting journalists, to learning how to take advantage of opportunities in new media.

At the same time the number of specialist science journalists has fallen sharply, largely because of the diminishing resources available to media organizations as their circulations and advertising revenues have shrunk. This has forced scientists to look for new channels of communication.

The emergence and exponential growth of the web is a powerful influence. It relies on a different style of writing and publishing, and takes the journalist-interpreter out of the equation. Social media has been transformational, with scientists seeing Twitter and blogging in particular as useful tools for communication. Mobile devices such as phones and tablets have become universal, and the improved quality of photography has added to their utility, especially when combined with simple systems for editing video footage.

The focus of research organizations has changed. In Australia there is more emphasis on commercial and industrial outputs from science, with governments and funding bodies expecting a return for their investment in research in terms of jobs, new industries and an improved economy. Keeping the public informed of their work has become a lower priority. The metrics used to judge researchers and research organizations revolve around the number of papers they have published and the number of times they are cited, and the rewards for public communication are scant.

We have responded to these changes by amending existing courses and adding new ones. We train scientists and journalists to take effective video on their smart devices and to edit it, and offer workshops for PhD students on writing for publication in scientific publications. The more traditional courses have been extensively modified to maintain their currency. It is an iterative and ongoing process of learning, testing and adapting our workshops to the needs of our participants.

Notes

1 For a full list of workshops available, see www.econnect.com.au/workshops/ (accessed June 26, 2018).
2 Science Minister Ross Free. Speech to ANZAAS Conference, Geelong August 1991
3 Science Minister Chris Schacht. Speech in presentation of Michael Daley Awards, Canberra April 1993. Quoted in article in *Sydney Morning Herald* by Peter Pockley.
4 Science Minister Barry O Jones. Speech at National Science Forum, CSIRO Limestone Avenue, August 1984, where he blamed scientists for not giving him the political support he needed in budget discussions.
5 John Stocker. *CoResearch* CSIRO internal newsletter October 1990.

6 Cribb, Julian Unpublished speech delivered to participants in CSIRO media skills course, 1992.
7 Dr. Gary Cook, CSIRO's Division of Wildlife and Ecology. Article in *CoResearch*, CSIRO internal newsletter, 1996–1997.
8 See Gascoigne, T. H. & Metcalfe, J. E. (1997). Incentives and impediments to scientists communicating through the media. *Science Communication*, *18*(3), March, 1997.
9 See (for instance) http://gunning-fog-index.com/ (accessed June 22, 2018).
10 Cathy Reade. Email to authors, July 20, 2011.
11 An example: Pose a simple question ("how many newspapers are there in the Pacific?") and give one person the portable microphone. Once they have guessed an answer, they pass the mike to the next person, and then along the row.

References

Barns, I. (1989). *Interpreting media images of science and technology*, Media Information Australia, 54.

Gascoigne, T. H. & Metcalfe, J. E. (1994). Public communication of science and technology in Australia. In B. Schiele (Ed.), *When Science becomes Culture: World Survey of Scientific Culture (Proceedings 1)*. Montreal: University of Ottawa Press.

Gascoigne, T. H. (2006). Scientists engaging with the public. In Cheng, D. et al. (Eds.), *At the Human Scale: International Practice in Science Communication*. Beijing: Science Press.

Eckersley, R. M and Woodruff, B. J. (1984). *Public Perceptions of CSIRO: A Staff Viewpoint*, Report to the CSIRO Executive, 22 March 1984.

Eckersley R. (1987). *Australian Attitudes to Science and Technology and the Future*. Canberra: Australian Government Publishing Service.

Mathews, I. (1991). *How to use the media in Australia*. Penguin.

Metcalfe, J. E. & Gascoigne, T. H. (1995). Science journalism in Australia. *Public Understanding of Science*, *4*(4), 411–428.

Metcalfe, J. E. & Gascoigne, T. H. (2001). Media Skills Workshops: Breaking down the barriers between scientists and journalists. *Pantaneto* Issue 3, July 2001.

Nelkin, D. (1990). Selling science. *Physics Today*, *46,* November.

Shortland, M. & Gregory, J. (1991). *Communicating Science*. London: Longman Scientific and Technical.

6 The challenges of writing science

Tools for teaching and assessing written science communication

Tzipora Rakedzon

Introduction

Overall, this chapter will address a general background on the training and assessing of science communication in terms of written communication – and the difficulties hidden in this specific skill. Indeed, the main source of information for the public is the online media, which ultimately belongs to the medium of writing (National Science Board, 2016), and therefore great importance must be devoted to teaching scientists to write about their science in a popular and accessible manner. To this end, it is important to understand the tools and goals of assessment of writing in general, those for assessing the basis of STEM (academic writing), and those of science communication. The chapter will present the difficulties of writing and the shift from writing for academic to lay audiences; a survey of practices recommended by existing training programs; key aspects of writing assessment such as genre and jargon; types and examples of writing assessment, including rating scales and computerized programs; and practical tools to be used in assessment and research of science communication. The chapter concludes with a summary of suggestions when using existing – or creating new – science communication training and assessment tools.

The difficulties of teaching and assessing writing

Anderson (1985) says that "reading and writing are the mechanisms through which scientists accomplish [science]." Indeed, talking or writing, and communication in general, is "the essence science" (Garvey & Griffith, 1979). Communication skills enable the sharing and contribution of knowledge, and in turn, acceptance into the scientific community, which is crucial to a scientist's career. The main obstacle in these communications lies in the fact that writing is often considered the most difficult of the four basic language skills (reading, writing, speaking, and listening) to master (Hamp-Lyons & Heasley, 2006; Kroll, 1990). In the case of science, graduate students and scientists must learn the conventions of written academic communication. These conventions include, among others, learning the scientific article structure IMRAD (Introduction – Methods – Results – Discussion) and learning advanced writing in English, for many second language learners (L2). This requires training as writing cannot be mastered, even by graduate students, without instruction (Wellington, 2010).

The problem is compounded in science communication. To understand the problem requires a definition of science communication. Science communication has many definitions, but a relevant definition for scientists making the transition from academic writing to writing for the non-expert, comes from Mercer-Mapstone and Kuchel (2015), an adapted version of Burns, O'Connor, and Stocklmayer (2003). They state that science communication is "the process of translating complex science into language and concepts that are engaging and understandable to non-scientific audiences such as politicians, industry professionals, journalists, government, educators, business, and the lay public (p. 2)." In this chapter, I concentrate on this adaptation and shift that scientists must deal with each time they tailor their message to a lay audience. The chapter reviews a variety of tools that exist for advanced writing skills in English in general, as it is both the language of science and a prerequisite for effective communication, and science communication tools and training programs in detail.

The genre shift

The translation of complex science into language for non-scientific audiences takes place, for example, when adapting academic articles to create a popular science piece on the same subject. This process is often referred to as a *genre shift* (Fahnestock, 1986). Genre in writing refers to types of literary productions, each with a different objective. In this shift, writing style and content, for example, have been shown to change. Genre instruction helps demonstrate the similarities and differences across genres (Devitt, 2015). Devitt (2015) explains that using examples of contrasting genres in class to discuss similarities and differences in style and content could benefit students by creating awareness of the target genre.

In her seminal paper examining the genre shift between academic and popular science article, Fahnestock (1986) found several changes, including less emphasis on methods, and more emphasis on the implications of the research. She also found major style differences, including the use of less cautious language in popular science.

Therefore, several of the tools introduced in this chapter are based on the premises of two theories: genre theory and sociocultural theory. Genre theory and the genre approach to teaching assert that different genres develop in different ways. In the genre approach to teaching writing, a particular genre is introduced and analyzed, then exercises on relevant language forms are given, followed by a short text that is produced by students (Dudley-Evans, 1997). It has become a popular and dominant approach to teaching writing over past few decades (Cope & Kalantzis, 2014; Devitt, 2015; Hyland, 2002; Swales & Feak, 2012).

The connection of genre theory and the genre approach to a social purpose and context is naturally complemented by the sociocultural approach, which focuses on the process of initiation and assimilation into the scientific community language and conventions. This "enculturation" into the community language and conventions (Newton & Newton, 1998), including genre conventions, help embed the dialogue of a specific discourse community into its constituents (Hyland, 2015).

Graduate STEM students are far along the process of enculturation into the scientific community. In their academic writing, they learn the primary genre of academic discourse that enables them to communicate and produce new knowledge in their community. Adding a somewhat contrasting genre, popular science, enables them to expand their scientific discourse and audiences, and helps widen their communication skills as scientists. Therefore, tools that synthesize genre theory and sociocultural theory can aid in science communication training, as well as research on science communication training. In turn, this research can shed light on scientists' and STEM students' acquisition of the two genres of writing.

Science communication training for graduate students and scientists

Training scientists to write is based on scientists' integration into the scientific community. This discourse of scientists is based on academic jargon and style used in academic and technical presentations at conferences and academic articles in peer-reviewed journals. As this is key to promotion in the scientific community, communicating science to the public has been neglected. This allows scientists to maintain an accepted distance from the public, and their elite status (Peters, 2013). In many cases, scientists are discouraged from speaking to the public, and such communication may lead to negative attitudes from their peers (Dunwoody, Brossard, & Dudo, 2009; Dunwoody & Ryan, 1985; Martinez-Conde, 2016). Moreover, scientists may not even have the words with which to talk about science with non-experts, nor do they have experience or training in doing so. But this is changing, as evident in the goals stated in the guide to STEM education: graduates should have "the capacity to communicate, both orally and in written form, the significance and impact of a study or a body of work to all STEM professionals, other sectors that may utilize the results, and the public at large" (National Academies of Sciences, Engineering 2018, p. 107). Studies have also shown that graduate students are also interested in learning to communicate with non-experts, and today many universities require a lay summary as part of the PhD dissertation (National Academies of Sciences, Engineering 2018). Therefore, training for science communication should begin as graduate students and continue in teaching scientists how to clearly get their message across (Brownell, Price, & Steinman, 2013; COMPASSonline, 2013; Gray, Emerson, & MacKay, 2005; Warren, Weiss, Wolfe, Friedlander, & Lewenstein, 2007).

As such, science communication training programs and workshops are on the rise (Besley & Tanner, 2011; Carrada, 2006; COMPASSonline, 2013; Crone et al., 2011). The "Directory of Science Communication Courses and Programs"

(Atkinson, Deith, Masterson & Dunwoody, 2007) lists US-based programs, and the "European Guide to Science Journalism Training" lists EU programs (European Commission, 2010). Moreover, several national science communication training programs are described by grey literature. These include the Australian National Centre for the Public Awareness of Science at the Australian National University, which has offered programs to train scientists to become "skilled communicators" since 1996. In addition, the Leshner Leadership Institute for Public Engagement with Science, established in 2015 as part of the American Association for the Advancement of Science (American Association for the Advancement of Science, 2017), offers a website and webinars focusing on helping scientists in developing and conveying their messages to non-experts. The European Science Communication Network designed and executed (between 2005 and 2008) communication training workshops for mostly early-career scientists (Miller & Fahy, 2009). Furthermore, the university-based Alan Alda Center for Communicating Science has offered programs since 2009 for science graduate students and scientists. This includes workshops, conferences, and lectures, as well as university journalism courses.

There are several suggestions for designing science communication training in the literature (Baram-Tsabari & Lewenstein, 2013, 2016; Rakedzon & Baram-Tsabari, 2017a). This limited literature concentrates mostly on recommendations for science communication training based on existing science communication courses. For example, Warren et al. (2007) make three recommendations: meetings with the institutes' press relations office to become familiar with the process and needs of the PR, visiting and meeting with reporters in the media, and experiencing actual hands-on experience in class, for example, writing a press release. Crone et al. (2011) describe learning activities and skills in a science communication course for graduate students. They primarily recommend working on general communication skills. These skills include the consideration of different types of audiences, and the ability to explain scientific concepts and processes. Practice activities include an interview with peers in the class without overusing jargon, as well as writing online articles and short radio broadcasts for a lay audience. Bishop et al. (2014) also recommends hands on experience and practice in science communication training for graduate students. Their course used blog writing, recommended peer and professional editing feedback, and concentrated on audience. For example, they suggested their students think of friends and family when writing for the non-expert audience (Bishop et al., 2014, p. B). Others studies also suggest the hands-on approach, incorporating learning from experts, and writing practice (Heath et al., 2014; Wilk, Spindler, & Scherer, 2016).

The above programs describe their goals of helping scientists convey their messages to non-experts. They concentrate on hands-on discussion and practice, as well as core skills such as knowing the audience, writing and speaking clearly, and familiarity with media (e.g., social media, news media, science cafés, and science festivals). These science sites, however, do not present any assessment tools for science communication training in general, nor for writing science, specifically.

Assessment scales

An examination of the literature shows that few studies have investigated scientific writing at any level and assessment rubrics are lacking. Similarly, the weakest link in the pedagogy of science communication, is assessment: Shared assessment tools for science communication training are almost non-existent (Baram-Tsabari & Lewenstein, 2013, 2016; Rakedzon & Baram-Tsabari, 2017a).

One primary assessment tool is the use of rubrics, i.e., rating scales. Rubrics for assessing writing are often employed in the evaluation of L2 writers, and mostly for assessing essays at the high school and undergraduate level (Crusan, 2010; Knoch, 2009a, 2009b; Polio, 1997). At the academic level, existing rubrics are mostly used for large-scale, standardized tests. These tests ask students to write about predetermined topics and enable the university to determine student needs and level before enrolling in an academic program (Educational Testing Services, 2005; Knoch, 2009b).

Several rubrics and scales are used to rate large-scale undergraduate writing. These include the Test of English as a Foreign Language (TOEFL), the Test for English for Educational Purposes (TEEP), the ESL Composition Profile (Jacobs, Zingraf, Wormuth, Hartfiel, & Hughey, 1981), and the DELNA (Diagnostic English Language Needs Assessment). At the graduate level, the graduate record examination (GRE), assesses students applying to graduate school. The GRE has an Analytical Writing Measure that uses a 1–6 holistic scale to rate controlled tasks requiring candidates to analyze an issue or argument. Another test, the International English Language Testing System (IELTS), also has a writing section; it is also used for acceptance into higher education or professional positions. These and similar tests are mainly assessment based, though they can be used to guide students through the preparation process. For example, they direct students to focus on some writing aspects relevant to graduate students such as coherence/cohesion; however, these tests are insufficient in assessing highly technical work by advanced graduate students as they neglect aspects such as vocabulary, content, and genre. For example, they require an essay format, but often do not include scientific genres. The inadequacy of these ratings lies in the purpose of such writing tests: they primarily fulfill diagnostic purposes to screen students before entering degree programs or to evaluate their need for English academic writing courses.

Existing writing assessment rubrics are also based on computer rating systems. Computerized rating scales are used by both teachers and researchers, but are limited in level and scope (for a review see Shermis & Burstein, 2013). The most widely used programs, such as the Australian AST Scaling Test, evaluate proficiency in genres such as writing an argumentative essay (McCurry, 2010). Another program used for high school and undergraduate levels, E-rater, was developed by the Educational Testing Service. E-rater provides a holistic score on an essay. In addition, Coh-Metrix has been tested on expository or persuasive genres in undergraduate students (Aryadoust & Liu, 2015). It measures various aspects of comprehension and coherence, including lexicon, syntax, and discourse (McNamara, Graesser, McCarthy, & Cai, 2014). However, these tests are also not designed for training, but can be used in studies examining shorter texts. Furthermore, like the aforementioned large-scale tests, they are capable of assessing controlled topics (Liu, Rios, Heilman, Gerard, & Linn, 2016).

The aforementioned rubrics, programs and research have stressed the need for additional assessment tools that can be used for both training and research. However, it is clear there are few examples in the literature on rubric development and few examples analyzing students' progress in writing genres (Boettger, 2010; Crusan, 2010; Crusan, 2015; Dempsey, PytlikZillig, & Bruning, 2009; Yayli, 2011). Furthermore, assessing popular science writing at the university level has seldom been investigated (Baram-Tsabari & Lewenstein, 2013) and systematic evaluation of learning outcomes in training programs has not been conducted (Baram-Tsabari & Lewenstein, 2016) This highlights the need to evaluate the programs training future scientists good written science communication. Many studies do stress the need for local assessment (Adler-Kassner & Harrington, 2010; Bonanno & Jones, 2007; Huot, 2010; Pagano, Bernhardt, Reynolds, Williams, & McCurrie, 2008). However, the studies supporting local, undergraduate level assessment have not been adapted for graduate level assessment, and specifically, not for popular science writing assessment.

Assessing jargon and readability

One primary issue when translating a message from academic to lay audience is the use of jargon. Academic and scientific texts are often brimming with jargon: academic texts have between 5–22% technical vocabulary, with the higher percentages found in academic science texts, such as biology, physics, and computer science (Nation, 2001; Hyland & Tse, 2007). This amount of jargon in a text makes academic and scientific texts difficult for the non-expert. Research has found that adequate comprehension requires knowledge of 98% of the words in a text for comprehension (Hu & Nation, 2000) – unlikely for many scientific texts. Moreover, there is a growing trend, especially in the medical and legal fields, to use "clear communication" and "plain language". This can be seen in guidelines for online patient information (OPI) by the National Institute of Health (NIH), the American Medical Association, and the Department of Health & Human Services (Garcia, 2018).

It is therefore not surprising that the most basic guideline for communicating science to the lay public that is agreed upon is the avoidance of jargon. Overuse of jargon results from the norms of academic discourse and the lack of general vocabulary to describe scientific ideas and phenomena. Jargon use also results from the "curse of knowledge," the situation in which scientists do not remember that in the past they, too, did not have knowledge of such terminology (Heath & Heath, 2007). To help scientists and STEM graduate students limit their jargon use when communicating with the lay public requires a tool to objectively evaluate jargon use. However, the evaluation of jargon use in general, and of science communication workshops by students and scientists specifically, has also, unfortunately, been neglected.

There have been few attempts to objectively categorize jargon, and even fewer to automatically identify it. Most existing literature have created field-specific lists, for example in medicine (Chen & Ge 2007; Wang, Liang, & Ge, 2008), agriculture (Martínez, Beck, & Panza, 2009), chemistry (Valipouri & Nassaji, 2013), applied linguistics (Khani & Tazik, 2013; Vongpumivitch, Huang, & Chang, 2009), engineering (Hsu, 2014), law (Benson, 1984) and computer science (Hyland & Tse, 2007; Tse & Hyland, 2009). These lists were compiled to help English writing course instructors in preparing students for higher education literacy. Although attempts to automatically categorize jargon were made by Baram-Tsabari and Lewenstein (2013) and Sharon and Baram-Tsabari (2014), these attempts did not produce a user-friendly tool that is easily used by leaners or trainers.

Another key to assessing writing level and appropriateness to audience is readability. Readability indices look at sentence length and word length to estimate the difficulty of a text. There are dozens of such tests, but the most widely used is the Flesch-Kincaid readability tests, which estimates the grade level of texts. As texts for the public are generally suggested to be written at a sixth to eighth grade level (Garcia, 2018), these readability tests can be used to test whether a text is suitable for a general audience. One readability index that combines tests can be found on https://readable.io/. However, readability tests are not enough to be used in isolation for instruction as they "are narrowly based on surface-level linguistic features" and do not truly reflect reader interaction with, or understanding of, the text (Crossley, Greenfield, & McNamara, 2008, p. 475).

Newly developed rubrics and rating programs and their use in assessment, research and training of writing skills

One research project has created genre rubrics to assess graduate level scientific writing in academic and popular science genres (Baram-Tsabari & Lewenstein, 2016; Boettger, 2010; Rakedzon & Baram-Tsabari, 2017a). The scoring rubrics were based on empirical evidence from an Academic Writing Course syllabus and student outcomes as well as on course goals and guidelines from the literature on academic writing (Day & Gastel 2011; Swales & Feak, 2012; Rakedzon & Baram-Tsabari, 2017b). Rubric development was conducted according to stages of rubric development in the literature (Crusan, 2010).

Development of the rubric began with the goal of focusing on the primary genre of scientists, academic writing from the IMRAD genre of scientific papers and advanced English proficiency. The rubric included writing aspects most salient to the literature and the course syllabus, and underwent three rounds to ascertain which descriptors were also measurable in the framework of a 14-week course (for a full description of the rubric development see Rakedzon & Baram-Tsabari, 2017b). A popular science scoring rubric for a science communication intervention lesson was developed according to the main goals of the literature on popular science writing assessment (Baram-Tsabari & Lewenstein, 2013) and students weaknesses found in the pilot rounds. The same descriptors of advanced English proficiency were also assessed in this genre, as fulfilling genre structure is not enough to produce a clear text: a correct and adequate level of English is also necessary.

The final version of the rubric can be used to assess science communication in isolation, or in comparison with an academic written text; the English proficiency rubric is designed to be used on both genres. The academic genre descriptors evolved to comprise use of (1) an informative title, (2) IMRAD (introduction, methods, results, discussion) format, (3) specific verbs (verbs like *compare* and *evaluate* instead of phrasal verbs such as *look at)* and (4) concise language (avoiding wordiness). The popular science rubric, on the other hand, assesses five major components: (1) the use of a catchy title, (2) active voice, (3) inverted pyramid (bottom line then background) format, (4) journalistic format (answers the 6 Wh-questions: what, who, when, where, why, how) and (5) definition/explanation for jargon. The final general English proficiency rubric comprised: (1) correct use of sentence structure (mistakes in run-ons, parallelism or commas as well as being over-simplistic or lacking variety in sentence structure); (2) coherence/cohesion (mistakes in connectors, paragraph structure/flow, or choppy sentences); and (3) verb tense and form (mistakes in choice of tense, voice or form). Marking for all rubrics was based on a 1–4 scale (for the full rubric see Rakedzon & Baram-Tsabari, 2017b).

Research on the development and use of these genre rubrics was based on the *English Academic Writing course for graduate students* at the Technion (Rakedzon & Baram-Tsabari, 2017a, 2017b). The course is compulsory for all doctoral students and elective for master's students on a thesis track. The (then) 14-week course aims primarily to prepare graduate students for the demands of scientific writing as found in journals and conference papers. Teaching academic writing comprises language and structure of the typical IMRAD article. The academic writing course syllabus also includes advanced English proficiency incorporating, for example, English sentence structure and advanced grammar. Therefore, the research compared pre- and post-tests to assess STEM graduate students' progress. Comparison of the contrasting genres was conducted using students' outcomes in writing an abstract to represent the academic genres, and a popular science press release to represent the popular science outcomes. For popular science writing, the press release genre was chosen, as it was the appropriate genre used in university PR departments, and a practical length for a one-time intervention within an academic writing course. Pre- and post-tests asked students to:

> Please describe your research, its context and implications for (A) a general audience (with no science background) and (B) the academic community in 150–250 words each (you can pick a specific project in progress or research that has already been completed).

Results showed an overall positive effect of the academic writing course, with improvement in both popular science and academic writing in groups that received a popular science intervention lesson with a popular science writing task. Overall, no interference was found, i.e., adding the popular science genre did not hinder graduate students' improvement in the academic genres. By contrast, the control group, which did not participate in the course, did not show significant improvements in either genre. The findings of this study suggest that single genre interventions in a writing course can be used as a tool to enhance the skill of writing science communication among STEM graduate students.

This rubric was designed for short interventions or training sessions and can be expanded to include other aspects of science communication, such as use of analogy, humor or storytelling, and readability level, which may be covered in longer intervention or training sessions. Further research could also examine other levels of trainees, including higher levels such as active researchers and scientists, as well as lower levels such as undergraduate students.

Another rubric that offers assessment options for science communication include "12 core skills for effective science communication" for teaching undergraduates (Mercer-Mapstone & Kuchel, 2015). This list concentrates on aspects such as identifying the audience and purpose of the communication, and use of appropriate language and style elements of analogy, humor, and storytelling. Baram-Tsabari and Lewenstein (2013) also enumerated the expectations of communicating popular science writing. They state seven main criteria for successful communication, including clarity and language, content, knowledge organization, style, analogy, narrative, and dialogue. These studies and recommendations for assessment could easily be used in combination with the previously mentioned genre rubrics to cover other important aspects of genre organization or grammar which are problematic in the transfer from academic to lay audience.

In addition to the genre rubrics, which are designed to be evaluated by human raters, and expanded upon to be used and tailored to individual training use, computer ratings offer an important option for large scale, and, sometimes, self-assessment by the writer. To automate jargon identification, we developed the De-Jargonizer at www.scienceandpublic.com. This program can be used to assess general vocabulary and jargon use to help scientists and science communication trainers adapt their texts and messages to various audiences. The program color codes vocabulary in texts according to three levels: high frequency/common words, mid-frequency, and jargon – rare and technical words.

Research using the De-Jargonizer has examined (Rakedzon, Segev, Chapnik, Yosef, & Baram-Tsabari, 2017) graduate students' use of jargon in academic and popular science writing, to investigate whether they change their use of jargon in writing texts for non-experts following an academic writing course with a popular science writing intervention. In this research, the De-Jargonizer was not available for students to use when writing but was used to rate student outcomes. Results showed that, overall, jargon was a confusing issue in popular science writing for graduate students. On one hand, most students clearly used less jargon in their popular science writing versus their academic writing.

However, students still used too much jargon in their popular science writing for a general audience. A tool such as the De-Jargonizer for showing writers which vocabulary is problematic during the writing process would probably have alleviated this problem.

Research using the De-Jargonizer can be used not only to test texts by scientists that have been adapted to a general audience, but also communication between professionals, such as doctors and lawyers, and their patients or clients. The program can also be easily used to test texts used in science classrooms, and lectures by science teachers at all levels.

Summary

When looking for or designing training and assessment for science communication, it is important to remember that different skills require different tools. All tools aim to tailor one's message for the audience. Some tools give automated feedback on technical terms, other on readability and grade level of a text. Rubrics exist for presentations, for written texts, and for comparison of popular science to academic genres. Examples of training sessions are given both in the literature, and through universities and science organizations and associations. Existing tools provide a good beginning for training, while much still needs to be developed. A combination of these automated programs and human-rating rubrics can greatly aid in evaluating science communication training both by the trainer and the leaner. In the future, there is a need for automated evaluation of text genre and content, especially for long texts and advanced scientific topics, and when a topic is not controlled. This area is still highly limited and almost non-existent.

References

Adler-Kassner, L. & Harrington, S., (2010). Responsibility and Composition's Future in the Twenty-first Century: Reframing & Accountability. *College Composition and Communication, suppl. SPECIAL ISSUE: The Future of Rhetoric and Composition, 62*(1), 73–99.

American Association for the Advancement of Science. (2017). Center for public engagement with science and technology. Available at www.aaas.org/pes (accessed January 31, 2017).

Anderson, R.C. (1985). Role of the reader's schema in comprehension, learning and memory. In H. Singer & R. B. Ruddell (Eds.), *Theoretical models and processes of reading* (pp. 372–384). Newark, DE: International Reading Association.

Aryadoust, V. & Liu, S. (2015). Predicting EFL writing ability from levels of mental representation measured by Coh-Metrix: A structural equation modeling study. *Assessing Writing, 24*, 35–58.

Atkinson, Deith, B., Masterson, K., & Dunwoody, S. (2007). *Directory of Science Communication Courses and Programs*. Madison, WI: University of Wisconsin.

Baram-Tsabari, A. & Lewenstein, B. (2017). Science Communication Training: What are We Trying to Teach? *International Journal of Science Education, part B, 7*(3), 285–300.

Baram-Tsabari, A. & Lewenstein, B. V. (2013). *An Instrument for Assessing Scientists' Written Skills in Public Communication of Science*,

Baram-Tsabari, A. & Lewenstein, B. V. (2016). Assessment. In M. C. A. van der Sanden & M. J. de Vries *Science and Technology Education and Communication* (pp. 163–185). Rotterdam: SensePublishers.

Benson, R. W. (1984). The End of Legalese: The Game is Over. *NYU Rev. L. & Soc. Change*, *13*, 519–573.

Besley, J. C. & Tanner, A. H. (2011). What Science Communication Scholars Think About Training Scientists to Communicate. *Science Communication*, *33*(2), 239–263.

Bishop, L. M., Tillman, A. S., Geiger, F., Hynes, C., Klaper, R., Murphy, C., … Hamers, R. J. (2014). Enhancing Graduate Student Communication to General Audiences through Blogging about Nanotechnology and Sustainability. *Journal of Chemical Education*, *91*(10), 1600–1605.

Boettger, R. K. (2010). Rubric Use in Technical Communication: Exploring the Process of Creating Valid and Reliable Assessment Tools. *IEEE Transactions on Professional Communication*, *53*(1), 4–17.

Bonanno, H. & Jones, J. (2007). *The MASUS Procedure: Measuring the Academic Skills of University Students: a Diagnostic Assessment.* Sydney: University of Sydney, Learning Centre.

Brownell, S. E., Price, J. V & Steinman, L. (2013). Science Communication to the General Public: Why We Need to Teach Undergraduate and Graduate Students this Skill as Part of Their Formal Scientific Training. *Journal of undergraduate neuroscience education : JUNE : a publication of FUN, Faculty for Undergraduate Neuroscience*, *12*(1), E6–E10.

Burns, T. W., O'Connor, D. J. & Stocklmayer, S. M. (2003). Science Communication: A Contemporary Definition. *Public Understanding of Science*, *12*(2), 183–202.

Carrada, G. (2006). *Communicating Science: A Scientist's Survival Kit*. Luxembourg: Office for Official Publications of the European Communities.

Chen, Q. & Ge, G. (2007). A corpus-based lexical study on frequency and distribution of Coxhead's AWL word families in medical research articles (RAs). *English for Specific Purposes*, *26*(4), 502–514.

COMPASSonline (2013). GradSciComm Workshop Summary. Available at: www.scribd.com/doc/191901955/GradSciComm-Workshop-Summary (accessed January 9, 2014).

Cope, B. & Kalantzis, M., 2014. *The Powers of Literacy (RLE Edu I): A Genre Approach to Teaching Writing*. New York, NY: Routledge.

Crone, W. C., Dunwoody, S. L., Rediske, R. K., Ackerman, S. A., Zenner Petersen, G. M., & Yaros, R. A. (2011). Informal Science Education: A Practicum for Graduate Students. *Innovative Higher Education*, *36*(5), 291–304.

Crossley, S., Greenfield, J. & McNamara, D. (2008). Assessing Text Readability Using Cognitively Based Indices. *Tesol Quarterly*, *42*(3), 475–493.

Crusan, D. (2010). *Assessment in the Second Language Writing Classroom*. Ann Arbor, MI: University of Michigan Press.

Crusan, D., 2015. Dance, ten; looks, three: Why rubrics matter. *Assessing Writing*, *26*, 1–4.

Day, R. A., & Gastel, B. (2011). *How to write and publish a scientific paper*. Santa Barbara,. CA: Greenwood.

Dempsey, M. S., PytlikZillig, L. M. & Bruning, R. H. (2009). Helping preservice teachers learn to assess writing: Practice and feedback in a Web-based environment. *Assessing Writing*, *14*(1), 38–61.

Devitt, A. J. (2015). Genre performances: John Swales' Genre Analysis and rhetorical-linguistic genre studies. *Journal of English for Academic Purposes, 19*, 44–51.
Dudley-Evans, T. (1997). Genre Models for the Teaching of Academic Writing To Second Language Speakers: Advantages and Disadvantages. In T. Miller (Ed.), *Functional Approaches to Written Text: Classroom Applications.* Washington, DC: United States Information Agency.
Dunwoody, S., Brossard, D., & Dudo, A. (2009). Socialization or Rewards? Predicting U.S. Scientist-Media Interactions. *Journalism & Mass Communication Quarterly, 86*(2), 299–314.
Dunwoody, S. & Ryan, M. (1985). Scientific Barriers to the Popularization of Science in the Mass Media. *Journal of Communication, 35*(1), 26–42.
Educational Testing Services (2005). Test of English as a foreign language. *www.toefl.com.*
European Commission (2010). *European guide to science journalism training*, Brussels.
Fahnestock, J., (1986). Accommodating Science: The Rhetorical Life of Scientific Facts. *Written Communication, 3*(3), 275–296.
Garcia, J. (2018). Communicating Discovery-Based Research Results to the News: A Real-World Lesson in Science Communication for Undergraduate Students. *Journal of microbiology & biology education, 19*(1).
Garvey, W. D., & Griffith, B. C. (1979). Scientific communication as a social system. *Communication: The essence of science*, p. 148. New York, NY: Pergamon Press.
Gray, F. E., Emerson, L., & MacKay, B. (2005). Meeting the Demands of the Workplace: Science Students and Written Skills. *Journal of Science Education and Technology, 14*(4), 425–435.
Hamp-Lyons, L. & Heasley, B. (2006). *Study Writing: A Course in Written English for Academic Purposes*. Cambridge: Cambridge University Press.
Heath, C. & Heath, D. (2007). *Made to Stick: Why Some Ideas Survive and Others Die*. New York: Random House.
Heath, K. D., Bagley, E., Berkey, A. J. M., Birlenbach, D. M., Carr-Markell, M. K., Crawford, J. W., … Wesslen, C. J. (2014). Amplify the Signal: Graduate Training in Broader Impacts of Scientific Research. *BioScience, 64*(6), 517–523.
Hsu, W. (2014). Measuring the vocabulary load of engineering textbooks for EFL undergraduates. *English for Specific Purposes, 33*, 54–65.
Hu, M. & Nation, I. S. P. (2000). Vocabulary density and reading comprehension. *Reading in a Foreign Language, 23*, 403–430.
Huot, B. (2010). Toward a new theory of writing assessment. *College Composition and Communication, 47*(4), 549–566.
Hyland, K. (2002). 6. Genre: Language, context, and literacy. *Annual Review of Applied Linguistics, 22*, 113–135.
Hyland, K., (2015). Genre, discipline and identity. *Journal of English for Academic Purposes, 19*, 32–43.
Hyland, K. & Tse, P. (2007). Is There an "Academic Vocabulary"? *TESOL Quarterly, 41*(2), 235–253.
Jacobs, H. L., Zingraf, S. A., Wormuth, D. R., Hartfiel, V. F., & Hughey, J. B. (1981). *Testing ESL Composition: A Practical Approach.* Rowley, MA: Newbury House Publishers, Inc.
Khani, R. & Tazik, K. (2013). Towards the Development of an Academic Word List for Applied Linguistics Research Articles. *RELC Journal, 44*(2), 209–232.
Knoch, U. (2009a). Diagnostic assessment of writing: A comparison of two rating scales. *Language Testing, 26*(2), 275–304.

Knoch, U. (2009b). *Diagnostic Writing Assessment: The Development and Validation of a Rating Scale*. Frankfurt: Peter Lang.

Kroll, B. (1990). *Second Language Writing: Research Insights for the Classroom*. Cambridge: Cambridge University Press.

Liu, O. L., Rios, J. A. Heilman, M., Gerard, L., & Linn, M. C. (2016). Validation of automated scoring of science assessments. *Journal of Research in Science Teaching*, *53*(2), 215–233.

Martinez-Conde, S. (2016). Has Contemporary Academia Outgrown the Carl Sagan Effect? *Journal of Neuroscience*, *36*(7), 2077–2082.

Martínez, I. A., Beck, S. C., & Panza, C. B. (2009). Academic vocabulary in agriculture research articles: A corpus-based study. *English for Specific Purposes*, *28*(3), 183–198.

McCurry, D. (2010). Can machine scoring deal with broad and open writing tests as well as human readers? *Assessing Writing*, *15*(2), 118–129.

McNamara, D. S., Graesser, A. C., McCarthy, P. M., & Cai, Z. (2014). *Automated Evaluation of Text and Discourse with Coh-Metrix*. USA: Cambridge University Press.

Mercer-Mapstone, L. & Kuchel, L. (2015). Core Skills for Effective Science Communication: A Teaching Resource for Undergraduate Science Education. *International Journal of Science Education, Part B*, 1–21.

Miller, S. & Fahy, D. (2009). Can Science Communication Workshops Train Scientists for Reflexive Public Engagement?: The ESConet Experience. *Science Communication*, *31*(1), 116–126.

Nation, I. S. (2001). *Learning Vocabulary in Another Language*. New York: Cambridge University Press.

National Academies of Sciences, Engineering. (2018). *Graduate STEM Education in the 21st Century*. Washington, DC: The National Academies Press.

National Science Board. (2016). *Science And Engineering Indicators*. Available at https://nsf.gov/statistics/2016/nsb20161/#/

Newton, D. P. & Newton, L. D. (1998). Enculturation and Understanding: some differences between sixth formers' and graduates' conceptions of understanding in History and Science. *Teaching in Higher Education*, *3*(3), 339–363.

Pagano, N., Bernhardt, S. A., Reynolds, D., Williams, M., & McCurrie, M. K. (2008). An Inter-Institutional Model for College Writing Assessment. *College Composition and Communication*, *60*(2), 285–320.

Peters, H.P. (2013). Gap between science and media revisited: scientists as public communicators. *Proceedings of the National Academy of Sciences of the United States of America*, *110* Suppl, 14102–14109.

Polio, C. (1997). Measures of linguistic accuracy in second language writing research. *Language Learning*, (March), 101–143.

Rakedzon, T., Segev, E., Chapnik, N., Yosef, R., & Baram-Tsabari, A. (2017). Automatic jargon identifier for scientists engaging with the public and science communication educators. *PLoS ONE*, *12*(8): e0181742. https://doi.org/10.1371/journal.pone.0181742

Rakedzon, T. & Baram-Tsabari, A. (2017a). Assessing and improving L2 graduate students' popular science and academic writing in an academic writing course. *Educational Psychology*, *37*(1).

Rakedzon, T. & Baram-Tsabari, A. (2017b). To make a long story short: A rubric for assessing graduate students' academic and popular science writing skills. *Assessing Writing*, *32*.

Sharon, A. J. & Baram-Tsabari, A. (2014). Measuring mumbo jumbo: A preliminary quantification of the use of jargon in science communication. *Public understanding of science, 23*(5), 528–546.

Shermis, M.D. & Burstein, J., (2013). *Handbook of Automated Essay Evaluation: Current Applications and New Directions*. New York, NY: Routledge.

Swales, J. M., & Feak, C. B. (2012). *Academic Writing for Graduate Students: Essential tasks and skills, Third Edition*, Ann Arbor, MI: University of Michigan Press.

Tse, P. & Hyland, K. (2009). Discipline and Gender: Constructing Rhetorical Identity in Book Reviews. In *Academic Evaluation*. London: Palgrave Macmillan UK, 105–121.

Valipouri, L. & Nassaji, H. (2013). A corpus-based study of academic vocabulary in chemistry research articles. *Journal of English for Academic Purposes, 12*(4), 248–263.

Vongpumivitch, V., Huang, J., & Chang, Y.-C. (2009). Frequency analysis of the words in the Academic Word List (AWL) and non-AWL content words in applied linguistics research papers. *English for Specific Purposes, 28*(1), 33–41.

Wang, J., Liang, S. & Ge, G. (2008). Establishment of a Medical Academic Word List. *English for Specific Purposes, 27*(4), 442–458.

Warren, D. R., Weiss, M. S., Wolfe, D. W., Friedlander, B., & Lewenstein, B. (2007). Lessons from Science Communication Training. *Science, 316*(5828), 1122.

Wellington, J., (2010). More than a matter of cognition: an exploration of affective writing problems of post-graduate students and their possible solutions. *Teaching in Higher Education, 15*(2), 135–150.

Wilk, A., Spindler, M., & Scherer, H. (2016). Scholar Development: A Conceptual Guide for Outreach and Teaching. *NACTA Journal; Twin Falls, 60*(4), 385–397.

Yayli, D. (2011). From genre awareness to cross-genre awareness: A study in an EFL context. *Journal of English for Academic Purposes, 10*(3), 121–129.

7 Insights for designing science communication training from formal science education

Apply the mantra and be explicit

Louise Kuchel

At the heart of every effective science communicator is the mantra "know your audience and purpose; understand your context and genre," and for those dedicated to continual improvement, the practice of evaluating the effectiveness of their work. The same might be said of an effective teacher. But aside from qualified teachers in formal education (mostly pre-university), most people who teach or train others have not been trained in how to design learning activities, assessment or evaluation. This is true for many science communication trainers and university instructors. As previous authors have noted, the disciplines of science communication training and science education have much to offer each other, but interaction between these disciplines is uncommon (Baram-Tsabari & Osbourne, 2015; Baram-Tsabari & Lewenstein, 2017). This chapter draws on the commonalities between effective communication and effective teaching to provide some practical guidance that may benefit science communication trainers and researchers.

More specifically, this chapter discusses current and best practice (often not the same) for how to approach the design of science communication training curriculum, assessment, and learning overall. It is about teaching and learning science communication and the practices described allow it to be evaluated. The discussion centers on formal tertiary science programs (undergraduate and graduate), but is relevant to formal education for professional science communicators and to trainers providing professional development to career scientists or communicators (e.g., workshops of one day to one week duration). This chapter stems from observations that basic concepts from formal education, such as setting clear learning goals and proper assessment of the attainment of those goals, are often missing from science communication training (Mercer-Mapstone & Kuchel, 2015; Baram-Tsabari & Lewenstein, 2017; Stevens, Mills, & Kuchel, 2019). I use the rhetorical situation as a framework for the discussion to help highlight the commonalities with best practice in communicating science. The information is grounded in research from education and illustrated using examples from science communication research. In doing so, my aims are to help bridge the gap between science communication and science education research and practice, and to enable people from each discipline to become familiar with key terminology and concepts from the other.

The article begins by discussing what science communication and training mean, provides some context to explain the growing interest in science communication training, and introduces how learning happens: In the section, *Know your context*, I briefly discuss different types of scientists and implications for science communication training in general. The next section, *Know your audience,* provides details about best and current practice in learning design and assessment which allow evaluation and quality enhancement of a program. I end with the sections, *Understand your purpose and tailor your genre* and *Be explicit*. I write this article from the perspective of my experiences as a scientist, a university educator, and researcher who has an active interest in finding practical ways to improve the ability of scientists to communicate effectively.

Know your context

What is science communication?

The words "science communication" conjure up a very different picture for different people. For science researchers, the words elicit images of writing scientific research papers or speaking at community events with the "general public" or being active on twitter, but rarely all three. They almost never mention "engagement" or interpersonal skills or casual conversations about science with friends and family. Professional science communicators respond with thoughts of activities that stimulate awareness, enjoyment, and interest about science (contemporary definition; Burns, O'Connor, & Stocklmayer, 2003), or regular two-way dialogue (e.g., Borchelt & Hudson, 2008), usually with non-scientists and rarely in formal education settings (e.g., the "ecosystem" model in Davies & Horst, (2016)). Teachers of science typically imagine written and oral works that can be graded, dialogue among peers to help them assimilate knowledge, and sometimes development of teamwork skills (Herok, Chuck, & Millar, 2013). The emphasis in response to these words differs between individual people, professional disciplines, and even geographical location. Importantly, enabling people to communicate in each of the ways described above requires different skill and mind sets. So it is important for science communication trainers and researchers to clarify a shared meaning for "science communication" with their stakeholders and readers, respectively.

Why the growing interest in science communication training now?

The professional expectations for communication by scientists are currently undergoing an evolution, some might say a revolution. There is the well-recognized need to make complex science more accessible to the public, in order to have a society well-informed on the issues that affect us all, such as climate change, human health and food supply, and scientists need to play a role in this. Scientists need to have a voice in the fast-paced interactive nature of social media and the internet to help counter fake news and misinformation, as well as to develop trustworthy and credible profiles. They are being encouraged to play a role in politics and policy making, and to collaborate with multidisciplinary, specialist scientific and non-scientific groups. These recent changes are additions to the changing expectations from within the scientific community, the disciplinary norms, for communicating science with their peers.

Among research scientists, there is active interest, discussion and debate about how to make increasingly specialized and complex science more accessible to other scientists. Almost every major journal in each discipline contains recent publications on this topic, including the prestigious journal *Nature* (Gewin, 2018). The discussions extend from how data is visualized and reported, to calls for more creative scientific writing. They include changes in writing style from passive to active voice and from third to first person, and to publication of databases and computer coding used in analyses. There has also been a rapid evolution in the way scientists communicate with one another and learn about emerging science and scientific practices with an explosion of blogs and online communities; most research groups will participate in at least one. The disciplinary norms for communication among scientists are changing, and this has significant implications for the formal education of scientists.

What is science communication training?

Likewise, the phrase "science communication training" elicits a variety of meanings. The term "training" tends to be used interchangeably with "education," despite there being a philosophical distinction between the two (Rickman, 2004). The most likely reason is that the learning goals of science communication training programs usually focus on developing and practicing skills (Baram-Tsabari & Lewenstein, 2017), which aligns most closely with the term training. Science communication training can refer to part of a formal education to become a scientist or a professional communicator. It may be a one-off event or a series of workshops that prepare scientists to engage with the journalists, policy makers or community groups. It may be informal or online, a duration of one hour or multiple days, and include non-scientists who volunteer in museums or participate in science events. It may involve theatrical techniques, practical application, role-playing or theory and values. Baram-Tsabari and Lewenstein (2017) provide a detailed overview of the variety on offer with reference to examples described in the research literature. I use the term science communication training instead of education throughout this chapter for consistency with the book in which it sits, but in tertiary education settings I view it as including skills, ways of thinking, theory and concepts, personal identity, values and attitudes (Rickman, 2004).

It is useful to draw a distinction between science communication training that takes place in formal tertiary education (undergraduate and graduate programs) and as professional development and as independent events, primarily because the goals and constraints are different. It is also important to distinguish between training for scientists and training for professional communicators. Formal education for scientists is primarily about learning how to do science and think scientifically; communication is one aspect of that. Formal education for professional

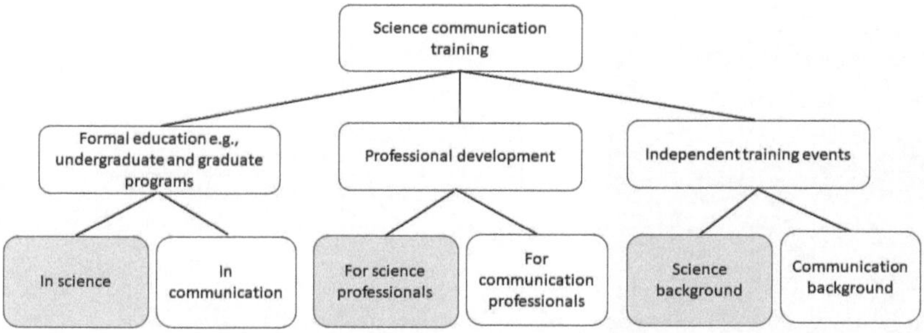

Figure 7.1 A schematic for the different settings in which science communication takes place. The purpose scope and design of training will different between these contexts.

communicators is dedicated to communication. The distinction between these contexts and those described above are often not made in the research literature, which creates confusion. It is important to clarify these distinctions if we are to advance the discipline and enable integration across disciplinary boundaries (Figure 7.1).

How learning happens

Despite the variety of contexts described above, all science communication training involves teaching for learning, and to be effective, how learning happens should be at the heart of the design of training events and programs.

Just like the scholarly discipline of communication, education theories and practices rise and fall in popularity and relevance, and are accompanied by unique jargon. I am going to focus on a few key ideas which are supported by scientific evidence and have practical implications for how to think about designing activities and assessment to promote learning. The descriptions I provide are in plain language, but I have provided key technical terms for readers interested in following up on the topic. I will revisit these ideas in the sections, *Understand your purpose* and *Be explicit*, where I discuss how they apply to best practice in learning design and assessment.

As most of us are aware, different people learn in different ways. (If the idea of visual, auditory and kinaesthetic learning styles springs to mind, please disregard. There is no evidence to support that people learn using preferred learning styles (Weale, 2017; Hood et al., 2017)). There is however a suite factors involved in everybody's experience of learning. The way these factors interact with each other is different for every person. The factors include (adapted from Race, 2014; Figure 7.2):

1 Motivation – intrinsic and extrinsic, or wanting and needing to learn.
2 Practice – learning by doing, verbalizing, experimenting, having a go, making sense of things, teaching, explaining, etc.

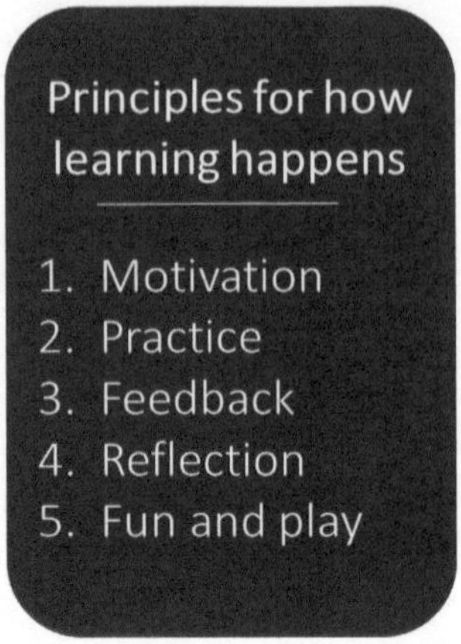

Figure 7.2 Principles that make learning happen.
Source: adapted from Race, 2014.

3 Feedback – learning from the opinions of other people, seeing results, praise, criticism, etc.
4 Reflection – making judgements such as comparing with criteria, deciding on strengths and weaknesses, self-assessment, judging the work of others. These all involve the internalization of criteria so that the learner automatically tries to meet them when performing a task.
5 Fun and play – this spans the four factors above; If we enjoy doing something, we are more likely to return to it and spend more time doing it.

To help people learn, each factor should be built into the design of any activity intended to help others learn. Each factor can be addressed one or more times, but the usual recommendation is to spend most time on practice and feedback (Race, 2014, 2015).

The reason that practice and reflection are key factors in making learning happen is that people construct meaning from what they do. In other words when learners practice, they link new knowledge and experiences with their existing knowledge and experiences. Reflection allows the learner to extrapolate this new knowledge to possible future scenarios and encourages the ability to transfer learning to other situations. In education this idea is known as constructivism (Piaget, 1971; Shuell, 1986; Semple, 2000, p. 25). Professional communicators are already implicitly aware of constructivism when they design communication to link old and new information and experiences and when they consider the values and beliefs of their audience.

Likewise, thinking about what the learner, or audience, does in response to the design of training should feel familiar to science communicators. In a constructivist view of learning, what the learner does is more important than what the teacher does (Shuell, 1986). This is often called a learner-centered (or student centered) approach to learning and it positions the teacher as a facilitator or coach rather than an instructor. Most people new to teaching spend a lot of their time and effort thinking about what they will do as the teacher: What information they will present, how they will explain things, and so on. This approach is typical of what educationists call direct instruction, a subsidiary of explicit instruction (more on explicit instruction in the section, *Be explicit*). In science communication training there are situations and circumstances where direct or explicit instruction is useful on its own or in conjunction with a student-centered approach, just as both the "deficit" and "dialogue" models are useful in science communication (Davies & Selin, 2012; Bauer, 2014).

Finally, communication is complex. Ample research demonstrates that teaching which emphasizes process over skills or tools is far more effective for complex learning. For example, research shows that when teaching writing, focusing on grammar only improves the writing of students already good at grammar (Braddock, Lloyd-Jones, & Schoer, 1963), whereas focusing on the rhetorical situation and the processes involved in conceptualizing, applying and refining writing improves the writing of all students (Bean, 2011). According to the review of the learning goals for science communication training by Baram-Tsabari and Lewenstein (2017), the majority of programs focus on skills development. I support their call for a broadening of goals addressed, especially goals and teaching approaches that focus on processes.

Know your audience

What is a scientist?

Much science communication training is aimed at scientists. A popular misconception is that all scientists are researchers in laboratories making new discoveries. This is far from reality and an important point for science communication trainers to consider when thinking about the goals, design and outcomes of their programs. A simple but useful descriptor of the various types of scientists in the workplace is the "10 types of scientist," produced by the Science Council in the UK (https://sciencecouncil.org/about-science/10-types-of-scientist/ for full description). The description is intended to increase career awareness for science students and is not exhaustive, but it may provide a useful starting point. The situations in which each type of scientist needs to communicate, the audiences, purposes, genres, best practice and conventions, all differ, as do their values, priorities and level of comfort with different types of situations. The discipline in which a scientist operates also influences the way scientists think, their values and priorities, and the norms in which they communicate (Fischhoff, 2018). Hence, it is important for science communication trainers to clarify and understand their target audience when designing training.

A broader concept of scientist needs to be considered by anyone designing training for formal science education, whether at undergraduate or graduate levels. This is because not every graduate of a science degree goes on to a career in science, and this is especially true for general science degrees and graduate research degrees (research Masters and PhD). For example, in Australia and the UK a minority of graduates from science Bachelor degrees are employed in traditional science occupations (Logan, et al., 2016; Palmer, Campbell, Johnson, & West, 2018). A study of PhD students in the U.S. indicates that around 40% are not interested in careers in academia (Roach & Sauermann, 2017) and there are recent calls for better data on career interests of graduate students (National Academy of Sciences, National Academy of Engineering, & Institute of Medicine, 2014). Graduates who do not go on to careers in science are scientifically literate people who apply scientific thinking to their work and lives, have interests in scientific issues and activities, who discuss and give advice on scientific topics to friends and family and more (Harris, 2012); an important subset to consider. Understanding the diversity of career destinations for learners you teach is an important part of deciding on learning goals for science communication training in formal education.

Understand your purpose and tailor your genre

Strategies for designing and evaluating curricula

Curriculum refers to a series or group of subjects that comprise a course of study. It is tempting to see an opportunity to teach communication within a curriculum and decide what to teach based on content, available expertise and resources. Conventionally, this is how university education was designed, and in some cases and many countries it still is (DeBoer, 2011; Biggs, 2014). This approach tends to lead to content-heavy, sometimes crowded programs, often providing limited practice in deep thinking and transferring knowledge and skills across contexts (Wiggins & McTighe, 1998), and fragmented tertiary curricula (Huber, 1992; Candy, Crebert, & O'Leary, 1994; Knight, 2001). To ensure teaching remains focused and organized, current best practice in education advocates using outcomes-based strategies to design training programs i.e., starting your design with the end in mind.

Backwards design (Wiggins & McTighe, 1998, 2006) and constructive alignment (Biggs, 1999; Biggs & Tang, 2011) are two outcomes-based strategies for designing curriculum and assessment (Figure 7.3). Designing curricula using these strategies begins with identifying desired results (learning goals and measurable learning outcomes). Once outcomes are identified, designers consider how these will be evidenced (assessment), and how learners can experience and practice what they learn (learning activities). Taking this approach helps to ensure teaching and learning remains focused and organized. Constructive alignment adds the idea that the learning activities involve learners in practicing the knowledge and skills required to have the best chance of success in the assessment (it is based on the notion of constructivism or learning by doing). The idea is that "the learner is 'trapped,' and cannot escape without learning what is intended." (Biggs, 2003, p. 27).

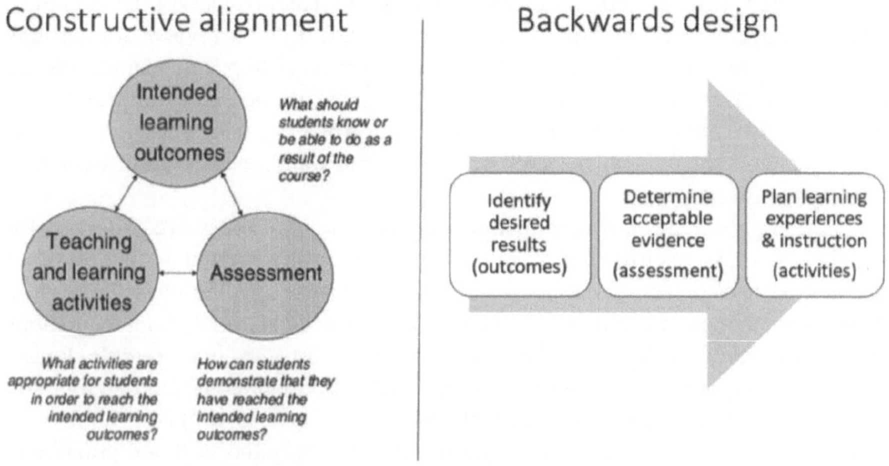

Figure 7.3 Illustrations of the steps involved in two outcomes-based strategies for designing curricula: constructive alignment and backwards design.

Both of these approaches enable evaluation of curricula at program and institutional scales by examining how well the learning goals, assessment tasks and learning activities align. This type of evaluation is called design-focused evaluation. To maximize the rigor of the evaluation, data should be collected and triangulated from three different sources; typically, these will include the student perspective, student performance against learning goals, and teacher perspective (Kember, 2003). Smith (2008) provides practical guidance on how to construct design evaluations from a student perspective, and Mercer-Mapstone and Kuchel (2015) is an example of design evaluation from a teaching perspective, while Matthews and Mercer-Mapstone (2018) is a combination of the two. An evaluation that examines student performance would use data from how well students perform on different assessment tasks or assessment criteria. These same sources of data can be used to revise and improve existing curricula using the framework by Wieman, Perkins and Gilbert (2010) which asks what should students learn? What are they currently learning? and what instructional approaches improve student learning? Biggs (2014) argues that designing and evaluating curricula using constructive alignment enhances the quality of educational programs and is more effective than quality assurance approaches (Biggs, 2014).

How to think about learning goals

So if we begin at the end, how do we go about identifying the goals and desired outcomes for science communication training? Most people will jump straight to thinking about what they want the students to know and to be able to do. But educationists identify learning more broadly. Most educational theorists describe learning in three domains: thinking and knowing (the cognitive domain), feeling and emotion (the affective domain) and doing (the psychomotor domain), (Bloom, Engelhart, Furst, Hill, & Krahtwohl, 1956). In their review of programs that address learning science in informal environments (e.g., museums, science events, etc.), Baram-Tsabari and Lewenstein (2017) provide an example of what this looks like, and Besley Dudo, Yuan, and Abi Ghannam (2016) discusses goals for workshops with research scientists that apply each without explicitly naming the domains.

Learning goals are in part shaped by the constraints of the genre or format for training. It makes sense that a short workshop not associated with any other training focuses on how to apply a specific skill or use specific technology. It is difficult to achieve more in such a short timeframe. Goals for longer workshops and for formal education have the opportunity to be broader. For example, longer workshops could consider including goals that encourage continued application of the skills, such as participants identifying themselves as confident communicators, or participants recognizing opportunities to apply the skill and practice in transferring the skill to different situations.

Identifying goals in formal education is more complex. The goals for science communication training need to align with the goals of the institution (university, college, etc.), the overall goals of the program (science or communication), and the expectations of relevant careers or professions. As such, the goals for training professional communicators will be very different to those for training scientists. In both cases however the goals will include disciplinary norms. One of the big challenges for formal education is that disciplinary norms and professional expectations are changing rapidly, and in many cases are quite different to what is currently being taught.

What are the goals for science communication training in formal science education?

In countries where outcomes-based thinking is applied in higher education, most tertiary institutions or governing education bodies will state the learning goals for both undergraduate and graduate science programs. They appear under various names, such as graduate attributes (UK and Australia), threshold learning outcomes (Australia), benchmarking statements (UK), core competencies and learning goals (US). They usually appear as a list of six or so statements that relate to science in general or sub-disciplines thereof, and they always include a statement about communication.

Many learning goal or outcome statements about communication are limited in how they conceptualize communication and in their usefulness for informing design. For example, the meaning may be too general, such as "Effective communication." They often only articulate outward communication, such as writing and speaking, and do not mention receiving communication, such as reading and listening. Many describe writing and speaking as products (Harper & Orr Vered, 2017), which ignores the value of learning the processes required to create them, the value of two-way communication, and normative communication practices in science such as clearly annotating databases and computer code, and so on. Learning goals for the units of study are more specific but typically reflect the limitations of the broader statements for the programs in which they sit. An exception tends to be learning goals for vocationally oriented programs. This comes about through oversight from accrediting or industry bodies and clear and specific understanding of the types of communication required by the profession e.g., medical practitioners, health and veterinary practitioners. These programs provide some excellent examples of learning goal statements and translation into professionally relevant learning activities (e.g., Herok et al., 2013; Skye, Wagenschutz, Steiger, & Kumagai, 2014; Barker, Fejzic, & Mak, 2018).

A growing number of learning goal statements for science programs do explicitly recognize that science students should learn to communicate with "broader audiences" or "lay audiences," and occasionally recognize other aspects of the rhetorical situation as well e.g., "… graduates will … be effective communicators of science by communicating scientific results, information, or arguments to a range of audiences, for a range of purposes, and using a variety of modes" (Australian national learning outcome for undergraduate science degrees, cited in Jones, Yates, & Kelder, 2011). This is a useful step towards recognizing changing professional expectations of scientists and diversity of career destinations but there is limited discussion in the science education or science communication research literature about the scope of what these broader statements mean (e.g., what types of audiences, modes and purposes?) and how it can be taught effectively in undergraduate and graduate science programs where there is limited time, expertise and competing priorities.

The discussion that does exist focuses on undergraduate education in general degrees but is limited in its scope and rigor. For example, Brownell, Price, and Steinman (2013) provide some ideas for how science communication targeting lay audiences can be incorporated into existing learning activities, but do not discuss overarching strategies for how these might be coordinated across a program of study. Harper and Orr Vered (2017) discuss the application of Writing Across the Curriculum (WAC) and Writing In the Disciplines (WID) approaches (popular in the US and UK) to teaching writing, both of which emphasize non-assessed writing practice with feedback to help students learn how to perform well in graded assessment tasks. But they stop short of describing how this might be applied to learning goals specific to science. Mercer-Mapstone and Kuchel (2017) identified core skills for effective science communication with non-technical audiences by consulting the research literature and practicing experts to help better define parts of the overarching learning goals, but did not discuss how these might be implemented in science. Promisingly, though, these core skills align well with those at the core of WAC and WID approaches. There is a need for increased research and rigorous discussion to identify the scope for communication, and models for pragmatically and effectively incorporating that scope into science programs.

Not surprisingly, then, there is a substantial gap between current practice and the ideals implied in learning goals which include broader audiences. For example,

Stevens et al. (2019) took the approach of looking at assessment tasks in Bachelor of Science programs at research intensive universities in Australia and how they aligned with the descriptors in the national learning outcome. They found that despite the national outcomes advocating broad contexts for communication, written and spoken assessment tasks addressed very narrow contexts and reflect traditional disciplinary norms (Figure 7.4). Brownell et al. (2013) noted in an analysis of undergraduate science curricula that none offered units of study focused on science communication to lay audiences despite it being a core competency for the program. The same pattern appears to be true for graduate science programs.

Several publications recognize that most graduate science programs are not preparing graduates for diverse career directions or modern professional expectations for research scientists (e.g., Kuehne et al., 2014; Roach & Sauermann, 2017). Most science communication training available to science graduate students is voluntary and must be sought out by the student. A few programs include a unit of study on developing science communication skills (e.g., Divan & Mason, 2016) and a couple of studies have loosely identified potential learning goals (e.g., Bray, France, & Gilbert, 2011; Kuehne et al., 2014). Fueled by the current volume of public outreach activities by research scientists and graduate students in the US, there is currently very active discussion about formally embedding science communication with the public into graduate

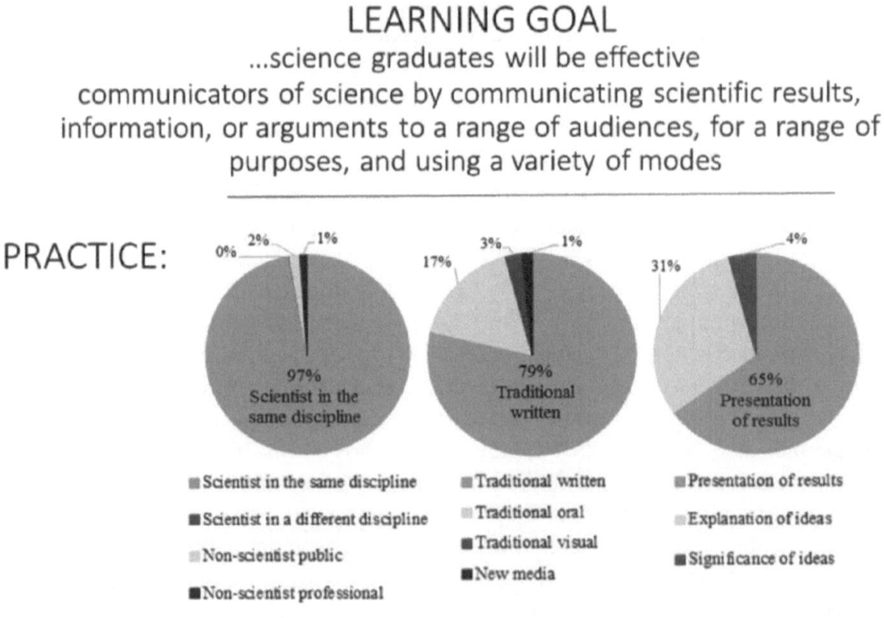

Figure 7.4 Evidence of the misalignment of assessment tasks (evidence of learning) with the Australian national learning goal for communication in science degrees using evaluation afforded by outcome-based strategies for designing curricula.

Source: adapted from Stevens et al (2019).

programs. For this to take place and be a meaningful addition to graduate outcomes, more research and discussion is required to identify the scope and translation of such goals into practice.

How to think about designing assessment

Everyone is familiar with the concept of assessment in formal education, for many seen as a form of punishment at the end of a course. So it is perhaps not surprising that the idea of assessing participants does not spring to mind when designing science communication training outside of formal education. We have already seen from several of the examples above that assessment within formal science education does not adequately align with learning goals for communication either, so it is worth briefly discussing the purpose and definitions of assessment. It may change the way you think about and use it.

At the heart of assessment is demonstrating what has been learned, and providing a measure of how well something has been learned. It is a form of simultaneous feedback and evaluation. It can be used by the learner to gauge how well they have attained a goal, and it can be used by the teacher to evaluate how well that goal has been taught. Assessment does not need to occur at the end of training, nor does it need to contribute to a formal mark or grade, though it can do both, and at the end of formal education it must demonstrate how well goals have been attained. Formative assessment (Cauley & McMillan, 2010) is a task that learners perform during the learning process, where they can self-evaluate or gain feedback on their strength and weaknesses in working towards a goal.

Instructors can use this information to immediately modify their teaching; either to address what learners are struggling with or, if learners are doing well, to move forward to the next step in the process. This use of assessment is very powerful and has been shown to improve learning and motivation. Formative assessment is very well suited to teaching process-driven, complex tasks such as science communication in all its forms, and is highly recommended for all forms of science communication training.

There are a few key features of an assessment task which help it to align well with a learning goal. The first is that the tasks best suited are authentic i.e., they are replicas of, or analogous to, tasks performed by professionals in the discipline (Wiggins, 1993, p. 229). The second is that the instructions describe how the task aligns with the goal (a rationale for the assignment). And third that the marking rubric includes criteria and/or statements that directly address the learning goal or associated, more specific, learning objectives.

How to think about designing learning activities

One of the most effective ways to begin thinking about learning activities is to identify the steps involved in the process of learning the knowledge or skill in question. Once this is done, it is easier to identify activities that will support or scaffold students in learning about and practicing these steps. Designing the activities to include the principles for learning outlined in the section, *Know your context,* and/or tasks for formative assessment will enhance the effectiveness of the activities.

Be explicit

Key to maximizing the learning that happens from using constructive alignment to design curricula is making the links between goals, assessment and learning activities visible and explicit to students (Biggs & Tang, 2011). This is especially so given that "assessment literally defines the curriculum for most students" (James & McInnes, 2001, p. 4). This can be achieved through explicit instruction which is "a structured, systematic, and effective methodology for teaching academic skills. It is called explicit because it is an unambiguous and direct approach to teaching" (Archer & Hughes, 2011, p. 1). Explicit instruction leaves nothing to chance and makes no assumptions about the skills and knowledge being taught (Torgesen, 2004, p. 363). It appears that making these links explicit for communication within formal science education is not common practice, as there have been many calls to make them more explicit (e.g., Brownell et al., 2013; Colthorpe, Rowland, & Leach, 2013; Herok et al., 2013; Jackson, Parkes, Harrison, & Stebbings, 2000; Oliver & Jorre de St Jorre, 2018)

Both explicit instruction and formative assessment require breaking down processes into steps and being explicit to students about both the steps and processes to which they lead. Often the knowledge about processes held by professionals within a discipline is tacit, and this poses a challenge for science communication trainers working with learners from a different discipline to their own. Randy Olsen's work (2009, 2015) is a nice example of an attempt to do this. He is a scientist who changed to a career in Hollywood filmmaking and transitioned back to working in science, mostly on science communication. He recognizes that scientists are analytical and have much more restricted time to spend on developing science communication thinking and skills, so has developed some easy-to-adopt, step-by-step processes to help scientists to apply strategies for creating stories and narrative. He is currently active in running "narrative circles" for scientists as a form of science communication training. Researchers from both science education and science communication training could contribute considerably to making tacit processes explicit by identifying and articulating the processes they use or struggle to understand in doing science communication.

Whilst explicit instruction may be an ideal fit for teaching communication in formal science education, current practice appears to be far from doing so. For example, Mercer-Mapstone and Kuchel (2015) compared the instructions and marking criteria for a sample undergraduate written assessment task against the core skills experts identified for communicating with non-technical audiences in undergraduate science (Mercer-Mapstone & Kuchel, 2017). They categorized each skill as explicit, implicit or absent. Perhaps not surprisingly, given what we have discussed already, most were absent, some were implicit and only a few were explicit (Figure 7.5). Interestingly more of the skills were explicit in

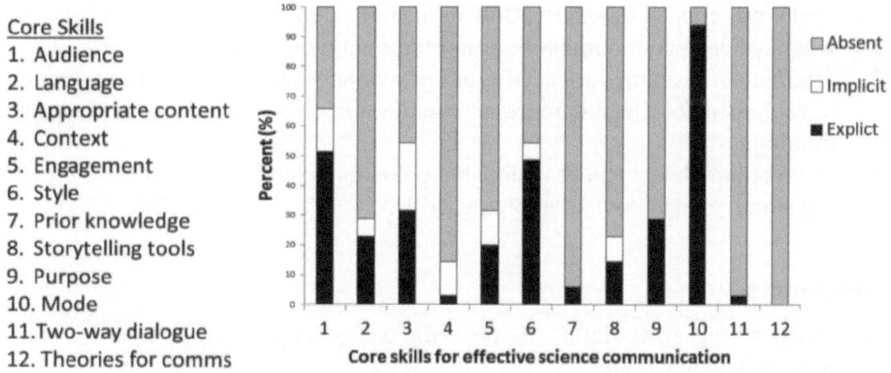

Figure 7.5 The appearance of core skills in instructions and marking criteria for a sample of 35 written assessment tasks from units of study in Australian Bachelor of Science programs. Seventeen tasks asked learners to write for non-technical audiences and 18 for technical science audiences.

assessment tasks that required communication to a non-technical audience. These findings highlight the enormity of the gap between current goals and practice for teaching "broader" science communication skills in many general science degrees. The liberal arts tradition of formal science education, where learners take courses such as rhetorical writing as part of their foundation subject, may have better outcomes for science students than programs that do not adhere to that approach.

Summary and conclusion

The intent of this chapter is the help identify commonalities between effective science communication and effective design for teaching it. It provides some practical guidance to science communication trainers by discussing best practices for, and examples of, current curriculum, assessment and evaluation design, using formal science education as the main context. In doing so, I have provided a framework that can be used at different scales by all science communication trainers operating both within and outside formal tertiary education to design, enhance, and evaluate training programs.

I have also highlighted a number of areas where collaboration among researchers from, and integration of, the disciplines of science communication training and science education would help to advance quality practices in these rapidly evolving and popular fields. These include:

1 Explicitly identifying tacit steps in complex communication processes; this is relevant to processes used by professional scientists and professional communications, and at all career stages from formal undergraduate education to seasoned professional;

2 Identifying ways in which those steps can be understood and operationalized with reduced time and effort for use by people (including scientists) for whom science communication is not their main occupation;
3 Finding and using a variety of types of evidence and frameworks to support development of clearer, more actionable goals for science communication training;
4 Developing clearer, more actionable goals for science communication training which extend beyond applying skills.

References

Archer, A., & Hughes, C. (2011). *Explicit Instruction: Effective and Efficient Teaching.* New York: Guilford Press.

Baram-Tsabari, A., & Lewenstein, B. (2017). Science communication training: what are we trying to teach?, *International Journal of Science Education, Part B, 7*(3), 285–300.

Baram-Tsabari, A., & Osbourne, J. (2015). Bridging science education and science communication research, *Journal of Research in Science Teaching, 52*(2), 135–144.

Barker, M., Fejzic, J., & Mak, A. S. (2018). Simulated learning for generic communication competency development: a case study of Australian post-graduate pharmacy students, *Higher Education Research & Development, 37*, 1109–123.

Bauer, M. W. (2014). A word from the editor on the special issue on 'public engagement', *Public Understanding of Science, 23*(1), 3–3.

Bean, J. (2011). *Engaging Ideas: the professor's guide to integrating writing, critical thinking and active learning in the classroom.* John Wiley and Sons, Incorporated.

Besley, J. C., Dudo, A. D., Yuan, S., & Abi Ghannam, N. (2016). Qualitative Interviews With Science Communication Trainers About Communication Objectives and Goals, *Science Communication, 38*(3), 356–381.

Biggs, J. (1999). *Teaching for Quality Learning at University.* Buckingham: Open University Press

Biggs, J. (2003). *Teaching for Quality Learning at University* (2nd ed.). Buckingham: SRHE and OUP.

Biggs, J. B. (2014). Constructive alignment in university teaching. *HERDSA Review of Higher Education, 1*, 5–22.

Biggs, J. B., & Tang, C. S. (2011). *Teaching for quality learning at university: what the student does.* Maidenhead: McGraw-Hill.

Bloom, B. S., Engelhart, M. D., Furst, E. J., Hill, W. H., & Krathwohl, D. R. (1956). *Taxonomy of Educational Objectives, Handbook I: The Cognitive Domain.* New York: David McKay Co Inc.

Borchelt, R., & Hudson, K. (2008). Engaging the scientific community with the public – communication as a dialogue, not a lecture. *Science Progress*, vol Spring-Summer, 78–81.

Braddock, R., Lloyd-Jones, R., & Schoer, L. (1963) *Research in Written Composition.* National Council of Teachers of English.

Bray, B., France, B., & Gilbert, J. K. (2011). Identifying the essential elements of effective science communication: What do the experts say? *International Journal of Science Education, Part B, 2*, 23–41.

Brownell, S. E., Price, J. V., & Steinman, L. (2013). Science Communication to the General Public: Why We Need to Teach Undergraduate and Graduate Students this

Skill as Part of Their Formal Scientific Training. *Journal of Undergraduate Neuroscience Education, 12*(1), e6–10.

Burns, T. W., O'Connor, D. J., & Stocklmayer, S. M. (2003). Science communication: a contemporary definition, *Public Understanding of Science, 12*, 183–202.

Candy, C. P., Crebert, R. G., & O'Leary, J. (1994). *Developing lifelong learners through undergraduate education.* Canberra: Australian Government Publishing Services.

Cauley, K. M., & McMillan, J. H. (2010). Formative Assessment Techniques to Support Student Motivation and Achievement. *The Clearing House: A Journal of Educational Strategies, Issues and Ideas, 83*(1), 1–6.

Colthorpe, K., Rowland, S., & Leach, J. (2013). *Good Practice Guide (Science) Threshold Learning Outcome 4: Communication.* Office for Teaching and Learning, Australian Government.

Davies, S. R., & Horst, M. (2016). *Science Communication: Culture, Identity and Citizenship.* London: Palgrave Macmillan.

Davies, S. R., & Selin, C. (2012). Energy futures: Five dilemmas of the practice of anticipatory governance. *Environmental Communication: A Journal of Nature and Culture, 6*(1), 119–136.

DeBoer, G. E. (2011). The globalization of science education. *Journal of Research in Science Teaching, 48*(6) 567–591.

Divan, A. & Mason, S. (2016). A program-wide framework to facilitate scientific communication skills development amongst biological sciences Masters students. Journal of Further and Higher Education, vol 40, no 4, pp. 543–567.

Fischhoff, B. (2018). Evaluating science communication. *Proceedings of the National Academy of Sciences*, Nov 2018.

Gewin, V. (2018). The write stuff: How to produce a first-class paper that will get published, stand out from the crowd and pull in plenty of readers. *Nature, 555*, 130.

Harper, R. & Orr Vered, K. (2017). Developing communication as a graduate outcome: using 'Writing Across the Curriculum' as a whole-of-institution approach to curriculum and pedagogy, *Higher Education Research & Development, 36*(4), 688–701.

Harris, K-L. (2012). *A background in science: what science means for Australian society.* Centre for the Study of Higher Education, for the Australian Council of Deans of Science.

Herok, G. H., Chuck, J., & Millar, T. J. (2013). Teaching and evaluating graduate attributes in science based disciplines, *Creative Education, 4*(7), 42.

Hood, B., Howard-Jones, P., Laurillard, D., Bishop, D., Coffield, F., Frith, U., … Foulsham, T. (2017). No evidence to back idea of learning styles. Letter to *Guardian*, 12 March, 2017. Retrieved from www.theguardian.com/education/2017/mar/12/no-evidence-to-back-idea-of-learning-styles

Huber, R. M. (1992). *How professors play the cat guarding the cream.* Fairfax, VA: George Mason University Press.

James, R., & McInnis, C. (2001). *Strategically re-positioning student assessment: A discussion paper on the assessment of student learning in universities.* Centre for the Study of Higher Education, The University of Melbourne. Available at: www.cshe.unimelb.edu.au

Jones, S., Yates, B., & Kelder, J. (2011). *Learning and teaching academic standards project: Science. Learning and teaching academic standards statement.* Sydney: Australian Learning and Teaching Council.

Kember, D. (2003). To Control or Not to Control: The question of whether experimental designs are appropriate for evaluating teaching innovations in higher education. *Assessment and Evaluation in Higher Education, 28*(1), 89–101.

Knight, P. T. (2001). Complexity and curriculum: A process approach to curriculum-making. *Teaching in Higher Education*, *6*(3) 369–381.

Kuehne, L. M., Twardochleb, L. A., Fritschie, K. J., Mims, M. C., Lawrence, D. J., Gibson, P. P., Stewart-Koster, B., & Olden, J. D. (2014). Practical science communication strategies for graduate students, *Conservation Biology*, *28*(5), 1225–1235.

Jackson N. J., Parks, G., Harrison, M., & Stebbings, C. (2000). Making the benchmarks explicit through programme specification, *Quality Assurance in Education*, *8*(4), 190–202.

Logan E., Prichard E., Ball C., Montgomery J., Grey B., Hill G., … , & Palmer K., (2016). *What do graduates do?* Higher Education Careers Services Unit, Prospects, Association of Graduate Careers Advisory Services, England.

Matthews, K. E., & Mercer-Mapstone, L. D. (2018). Toward curriculum convergence for graduate learning outcomes: academic intentions and student experiences, *Studies in Higher Education. 43*(4), 644–659.

Mercer-Mapstone, L. & Kuchel, L. (2015). Teaching Scientists to Communicate: Evidence-based assessment for undergraduate science education, *International Journal of Science Education*, *37*(10), 1613–1638.

Mercer-Mapstone, L., & Kuchel, L. (2017). Core Skills for Effective Science Communication: A Teaching Resource for Undergraduate Science Education, *International Journal of Science Education, Part B*, *7*(2), 181–201.

National Academy of Sciences, National Academy of Engineering, & Institute of Medicine (2014). *The Postdoctoral Experience Revisited.* Washington, DC: The National Academies Press.

Oliver, B., & Jorre de St Jorre, T. (2018). Graduate attributes for 2020 and beyond: recommendations for Australian higher education providers, *Higher Education and Research Development*, *37*(4), 821–836.

Olsen, R. (2009). *Don't be such a scientist.* USA: Island Press.

Olsen, R. (2015). *Houston we have a narrative: why science needs story.* USA: University of Chicago Press.

Palmer, S., Campbell, M., Johnson, E., & West, J. (2018). Occupational Outcomes for Bachelor of Science Graduates in Australia and Implications for Undergraduate Science Curricula, *Research in Science Education*, *48*(5), 989–1006.

Piaget, J. (1971). *Psychology and Epistemology: Towards a Theory of Knowledge.* Grossman: New York.

Race, P. (2014). *Making Learning Happen: 3rd edition.* London: Sage.

Race, P. (2015). *The Lecturer's Toolkit: 4th edition.* Abingdon, Routledge.

Rickman, P. (2004). Education versus training, *Philosophy Now: a magazine of ideas*, *47.*

Roach, M., & Sauermann, H. (2017). The declining interest in an academic career. *PLoS ONE*, 12(9), e0184130.

Semple, A. (2000). Learning theories and their influence on the development and use of educational technologies, *Australian Science Teachers Journal*, *46*(3), 21–22.

Shuell, T. J. (1986). Cognitive Conceptions of Learning. *Review of Educational Research*, *56*(4), 411–436.

Skye, E. P., Wagenschutz, H., Steiger, J. A., & Kumagai, A. K. (2014). Use of interactive theater and role play to develop medical students' skills in breaking bad news. *Journal of Cancer Education*, *29*(4), 704–708.

Smith, C. (2008). Design-focused evaluation. *Assessment & Evaluation in Higher Education*, *33*(6), 631–645.

Stevens, S., Mills, R., & Kuchel, L. (2019). Teaching communication in general science degrees: highly valued but missing the mark. *Assessment and Evaluation in Higher Education, 21*, 1–14. doi: 10.1080/02602938.2019.1578861

Torgesen, J. K. (2004). Lessons learned from research on interventions for students who have difficulty learning to read. In P. McCardle & V. Chhabra (Eds.), *The voice of evidence in reading research* (pp. 355–382). Baltimore, MD: Brookes.

Weale, S. (2017). Teachers must ditch 'neuromyth' of learning styles, say scientists, *The Guardian*, 13 March.

Wieman, C., Perkins, K., & Gilbert, S. (2010). Transforming science education at large research universities: a case study in progress. *Change: The Magazine of Higher Learning, 42*(2), 6–14.

Wiggins, G. P. (1993). *Assessing student performance*. San Francisco: Jossey-Bass Publishers.

Wiggins, G., & McTighe, J. (1998). What is backward design? In G. Wiggins & J. McTighe *Understanding by Design* 1st edition (pp. 7–19). Upper Saddle River, NJ: Merrill Prentice Hall.

Wiggins, G., & McTighe, J. (2006). *Understanding by Design.* Pearson: Merrill Prentice Hall.

8 Evaluating science communication training

Going beyond self-reports

Yael Barel-Ben David and Ayelet Baram-Tsabari

Introduction

A great deal of time, energy and experience is invested in science communication training programs (Salas, Tannenbaum, Kraiger, & Smith-Jentsch, 2012) and in supporting scientists to be better communicators. But there are very few measures to determine whether these programs are successful or in what ways they should be adjusted (Baram-Tsabari & Lewenstein, 2017a). Although the field of evaluation is applicable to a vast variety of professions and practices, few studies have examined evaluation in the context of science communication training programs. To determine effectiveness, training programs need to define specific goals and incorporate models of evaluation before, during, and after the intervention. This chapter reviews evaluation efforts in the field of science communication training and suggests what is needed. We define evaluation, what differentiates it from research and address its importance in our context; we also advocate for the importance of defining learning goals to conduct a meaningful evaluation. We then put forward evaluation approaches, methods and a framework originating in HRD to do so. Finally, we address the limitations on attempts to evaluate science communication programs.

Overview of science communication training program evaluations

Considerable research has been conducted in the past few years on scientists' views of science communication outreach programs and activities (Besley, Dudo, & Storksdieck, 2015; Ecklund, James, & Lincoln, 2012; Entradas & Bauer, 2016; Grand, Davies, Holliman, & Adams, 2015; Peters, 2012), outreach program effectiveness (Bogue, Shanahan, Marra, & Cady, 2013; Haran & Poliakof, 2011; Peterman, Robertson, Cloyd, & Besley, 2017; Sevian & Gonsalves, 2008) and factors affecting scientists' participation in these activities such as social norms and academic rank which were found to be corelated with scientists' engagement and outreach activities (Besley, 2015a; Besley, Dudo, Yuan, & Lawrence, 2018; Besley, Oh, & Nisbet, 2013; Cerrato, Daelli, Pertot, & Puccioni, 2018; Dudo, 2012; Dunwoody, Brossard, & Dudo, 2009; Poliakoff &

Webb, 2007; Robertson Evia, Peterman, Cloyd, & Besley, 2018; Royal Society, 2006).

Other studies about scientists' perceptions of science communication training programs have found that scientists acknowledge the importance of communicating their scientific work to the public and have positive attitudes toward science communication training programs (Besley, 2015b; Besley et al., 2015; Besley & Tanner, 2011; Besley & Nisbet, 2013; McCann, Cramer, & Taylor, 2015). Studies examining participants' and practitioners' objectives and goals indicate that training programs should provide guidance in choosing context-specific communication objectives in addition to teaching communication tools and skills (Dudo & Besley; Nisbet & Scheufele, 2009, 2016); and by mapping participants' and practitioners' priorities in science communication training (Besley, Dudo, Yuan, & Abi Ghannam, 2016) and pointing to key skills for effective science communication such as promoting trust, using clear and vivid language, tailoring the message to the audience, etc. (Baram-Tsabari & Lewenstein, 2012, 2017a; Brownell, Price, & Steinman, 2013; Mercer-Mapstone & Kuchel, 2015a, 2015b; Metcalfe & Gascoigne, 2009; Montgomery, 2017; National Academies of Sciences Engineering and Medicine, 2017; Rakedzon & Baram-Tsabari, 2016, 2017; Sevian & Gonsalves, 2008). These works have yielded ample data on scientists' reasons for communicating their science to the general public, their views on these endeavors and ideas about the core skills that can promote science communication. However, few studies have evaluated science communication training programs in terms of their effectiveness in improving participants' skills, their usefulness or impact (Cameron et al., 2013; Peterman et al., 2017; Rodgers et al., 2018; Rowland, Hardy, Colthorpe, Pedwell, & Kuchel, 2018). More substantial research and evaluation, addressing a broader range of science communication training programs skills and agendas, is needed to better guide professionals in ways to support scientists in communicating with the public effectively.

Evaluation and assessment approaches

Evaluation is sometimes defined as steps taken to judge the value of an action or an activity (Raizen & Rossi 1981).[1] This implies subjectivity, since value may differ from one person to another. The question of value for whom and for what purpose affects the way the results of evaluative actions are interpreted. Patton suggested that "evaluation involves making judgments about what is meaningful" (Patton, 2015, p. 5), and incorporated flexibility into the definition by stating that the meaning can vary between people or between stages of a program even for a given individual. The challenge of evaluation subjectivity, its goals and context-dependent characteristic make it difficult to generalize evaluations across programs.

Evaluation is not monolithic. The literature differentiates between formative, summative, process, impact and diagnostic evaluation, to name a few (Bates, 2004; Black, 1993; Black & Wiliam, 1998; Bogue et al., 2013; Chapelle, Cotos,

& Lee, 2015; Harlen & James, 1997; Khandker, 2010; Patton, 2015; Wiliam & Black 1996). Formative evaluation provides feedback to participants and instructors on current strengths and weaknesses (e.g., midterm quiz, questions asked during a class). Summative evaluation is mostly used at the end of the learning process and measures against a standard or validated benchmark to report an outcome (e.g., a test, a final project etc.). Diagnostic assessment takes place prior to a program and is used as an initial phase or as a baseline for participants' skills and knowledge.

At the program level, process evaluation is primarily used at the end. It focuses on program operation implementation and examines whether the program was executed as planned, what changes were made to the program over time, the reasons for the change and how the program process could be more efficient. Impact evaluation examines the entire program span in retrospect and measures the long-term effects of the program on policy or changes in behavior.

When assessing the effectiveness of training programs, a mixed evaluation approach should be used as a function of the goals one sets to achieve. Therefore, it is important to define desirable learning goals and outcomes for science communication training before evaluation is put into place.

Evaluation vs. research

Conducting an evaluation differs from doing research primarily in terms of its goals. Evaluation is aimed at examining the effectiveness of an intervention as a function of a specific goal and for the purpose of improving it (McGillin, 2003). The notion of evaluation incorporates an element of judgment of that action, which does not necessarily enter into research. Thus, evaluation is aimed at improving an intervention whereas research mainly involves epistemic goals such as understanding a phenomenon. Whereas research is used to advance theory and hence aims for generalization, evaluation has implications that are valid primarily for the program that is being assessed.

While in research it is essential to control variables and randomize samples (when possible), these are not usually operational when evaluating the effectiveness of an intervention. We cannot control the participants' characteristics or randomize them or the training program itself. For example, most science communication training programs cannot control parameters such as instructors or participants for the sake of research since they have other priorities, such as adjusting the program to the clients' needs or providing a coherent training product between training sessions. While research is dictated by the research questions, evaluation is dictated by the goal of improving the effectiveness of the intervention or to ensure superiors of the return on their investment.

Defining learning goals and outcomes

What constitutes a successful intervention? More specifically, what are we trying to teach and how can we know we have been successful? We need to clearly define learning goals and desired outcomes to initiate the evaluation of science communication training.

Baram-Tsabari and Lewenstein suggested building on concepts from science education to develop an assessment for performance in science communication (Baram-Tsabari & Lewenstein, 2016). They put forward a framework for the assessment of science communication training, based on informal science learning literature, that includes affective, content, methods, reflective, participatory and identity features (Baram-Tsabari & Lewenstein, 2017b). These very broad aspects of learning science communication can be used to develop specific learning goals when structuring a workshop, an intervention or a training program. Another potential starting point is the National Academy of Sciences Engineering and Medicine's (2017) five goals for science communication.[2] These can be used as scaffolding for thinking about the goals for teaching scientists to communicate.

Choosing learning goals is based on the specific program's agenda and philosophy: some programs are designed to teach how to communicate clearly in writing, whereas others are interested in trust building in face-to-face situations, or changing scientists' normative view as to the relationship between science and society. Whatever the primary goals are, they need to be developed into operative learning objectives that will guide the evaluation. For the primary goal of teaching how to communicate clearly in writing, measurable learning objectives could include reduced usage of jargon, making connections to the audience's day-to-day lives and the use of analogies. For the primary goal of supporting trust building in face-to-face situations, the measurable learning objectives could include incorporating personal stories or practicing active listening.

We think of science communication training as one type of professional development intervention. Structuring learning goals as part of the planning of a training program lays the groundwork for developing measurable learning outcomes and using relevant assessment tools and analysis processes for measuring its effectiveness.

From goals to tools – diverse ways to evaluate programs

Setting learning goals naturally influences the tools used for collecting and analyzing the data. For example, if the goal is to strengthen the participants' identity as science communicators, using focus groups to evaluate the participants' products will not provide exploitable information since it will yield an external judgment that will not necessarily assess inherent changes in the participants' sense of identity. Rather, a participant centered approach such as interviews or questionnaires would be more fitting. The nature of the goal drives the choice of questionnaire items or the observation rubric. For example, if the objective is to assess participants' writing skills, observing them pitch their research to investors, would not promote understanding of the program's impact on their writing, as appose to conducting a written task or evaluate participants based on portfolios.

Before selecting specific tools for specific outcomes, the general approach to evaluation must be shown to fit the purpose. If the program focuses on teaching skills, the evaluation approach should explore changes in competence as a result of participating in the program. Here we suggest an array of approaches, pragmatic in nature, and afterwards we divide them to self-reports and external evaluations.

Many intervention designs can be used for outcome evaluation, including a before and after array, comparison of different interventions, or different participants, or using a control group to evaluate program impact, to name a few (for examples see Bogue et al., 2013; McCann et al., 2015; Rodgers et al., 2018). Each design can also combine self and external reporting tools, as detailed below.

Self-reports

Self-report tools refer to any form of assessment asking participants to provide information about their own situation, attitudes or interpretation of a situation. Self-reports allow us to access information about participants that cannot be obtained by mere observation such as participants' perspectives and beliefs (Stone et al., 1999). Because of their potential, self-reporting tools are used abundantly in psychology, education, health studies and other social sciences to probe participants' opinions and perceptions about a topic (Brutus, Aguinis, & Wassmer, 2013). Although commonly used and recognized as providing valuable information, self-report tools have limitations. Participants answering a self-report survey may try to present a favorable image of themselves, or be dishonest, and thus skew the results. A different interpretation of a question than the questionnaire writers intended can also contribute to skewed results (Baldwin, 1999). Self-reports can be useful when evaluating the effectiveness of a program from the participants' point of view, for example, with regard to the affective, content, methods and identity goals suggested by Baram-Tsabari and Lewenstein (2017b). Surveys and interviews are two of the most common self-report techniques used in evaluations.

Surveys

Surveys are an effective way to collect information from a large number of people and can be analyzed quantitatively or qualitatively. Surveys are a method of assessment that are utilized widely for their simplicity and ability to access large numbers of respondents and gather substantial amounts of data. There are basically two kinds of questions used in surveys: close- and open-ended questions. In most close-ended questionnaires participants are asked to respond to a set of multiple-choice questions or position themselves on a Likert scale to rate their opinions, attitudes, satisfaction, etc. depending on the purpose of the survey. By contrast, an open-ended questionnaire asks participants to express their opinions, describe their experiences, etc. freely and in their own words (Tucker, 2014). Most surveys do not permit respondents to freely express themselves and limit the information to pre-defined questions. Although in research this is a desirable outcome, when conducting evaluations to measure participants' "take home message" a looser approach should be considered (Upcraft & Schuh, 2002). Close-ended and open-ended questions can be combined in a survey. Surveys can be distributed both online and off-line to substantial numbers of people and can be used in a pre-post layout.

Survey limitations: One of the key limitations is length in that an overly long questionnaire might affect the number of valid entries. Distribution can also be a limitation since if not administered online, distributing can be challenging, and not all potential recipients will receive the questionnaire, a problem that might result in a response bias. The return rate can range from 30% to 60%. This varies across disciplines, methods of administration, length and even number of reminders (Sheehan, 2006; Baruch & Holtom, 2008). Finally, constructing questions for a survey takes time and mastery to avoid biases.

Interviews

Interviews are a powerful tool in probing participants' experiences. An interview is mainly a qualitative research tool that allows the interviewee to share his or her perspective on an issue. There are numerous types of interviews aimed at different goals including narrative interviews, semi-structured interviews, stimulated recall interviews, structured situational interviews and others (Burden, Topping, & O'Halloran, 2015; Creswell, 2003; Lyle, 2003; Shubert & Meredith, 2015). Interviews permit a great deal of flexibility as regards themes and notions brought up by the interviewees and gives them control over turns in the conversation. A stimulated recall interview asks interviewees to reenact a process with an emphasis on sharing their line of thought when in the same situation. Interviews can provide insights into the interviewees' mindsets by asking participants how they accomplished a task such as explaining a complex scientific concept to a 10-year-old or presenting their research to investors before and after the intervention, why they changed things and how, etc.

Interview limitations: The number of participating interviewees can vary from several to dozens since interviews are very time consuming. Interviews, especially open-ended ones, vary greatly in length, depending on the goals and protocol but also on participants' willingness to talk at length. In addition, data analysis is more complex and demands control in qualitative analysis methods such as thematic inquiry and narrative analysis. These are skills that take time to master. In general, participants' willingness to participate affects the sample size and may not result in a representative sample.

External evaluation

External evaluation can be done by experts, peers, audience members, etc. Although this form of evaluation cannot yield information about participants' opinions or beliefs, it does provide an outside, "neutral" evaluation of the program's effectiveness and constitutes yet another way to evaluate the impact of training in real world settings. This form of evaluation can be implemented on several types of training products such as texts or videos of an individual participant. Since external evaluation is usually done on training products and evaluated against an assessment rubric or the evaluators' proficiency, it is best suited to evaluate skills. For example, in the fields of psychology and education, a skill development evaluation is often done by in-class simulations or by experts who observe participants implementing new skills (Kraiger, Ford, & Salis, 1993; Tucker 2014). This evaluation can be assigned to experts who assess changes in skills or their use in research and experience-based knowledge, by peers acquiring the same set of new skills and contributing from their perspective, or by lay audience members, who are the end clients in the case of scientists' training programs.

For example, in training programs that emphasize oral communication, formative evaluation such as peer-review can be implemented, since during most programs each participant has a chance to practice speaking skills in front of their peers at least once. The other participants can assess these talks using an evaluation sheet[3] that specifies the key categories and competencies addressed during the workshop. This evaluation can be done by all participants, thus helping them to develop a critical sense of what constitutes a good popular science talk, or by expert practitioners who can provide constructive feedback. This form of evaluation is employed to some extent in programs asking other participants to provide constructive feedback to fellow participants. Nevertheless, giving feedback is an art in and of itself and some instruction about how to give constructive feedback needs to be discussed beforehand. To improve evaluation and feedback, external evaluators of all types (peers, experts and lay audience) need to be provided with a strict and validated rubric to reduce cognitive workload and variability across raters and participants (Jonsson & Svingby, 2007).

Similarly, focus groups of experts or lay audiences can assess the pre- and post-output of participants in a training program by exposing these audiences to this output in a mixed, semi-blind order. Focus groups can provide a wide range of reactions and interpretations, thus enabling a glimpse into an authentic process of interaction with the scientist and the level of clarity, understanding and interest the intervention (Eliot & Associates, 2007; Nagle & Williams, 2004; Masadeh 2012). These types of external evaluation are suitable for written and oral material, including filmed videos or live observations (Rodgers et al., 2018).

Other types of external evaluation of intervention outcomes include using a computerized tool to evaluate level of reading difficulty such as the De-jargonizer – a free online automatic tool that measures the percentage of jargon in a text to help users adapt it to a lay audience (Rakedzon, Segev, Chapnik, Yosef, & Baram-Tsabari, 2017). A detailed description of this tool can be found in Chapter 6 of this book.

In general, it is preferable to use several validated and reliable instruments to evaluate the effectiveness of a training program to increase validity and strengthen the findings. Triangulated evaluation can be found in HR assessments where employees are asked to assess their own work which is then compared and contrasted to similar evaluations by their supervisors and/or colleagues

(Atkins & Wood 2002; Bracken, Timmreck, Fleenor, & Summers, 2001; Holt, Boehm-Davis, & Beaubien, 2001). A similar approach can be incorporated into science communication program evaluations.

A model for evaluation

Different evaluation models implemented in fields such as education, management and HRD can be adapted to the evaluation of science communication training programs. Below we present an evaluation model taken from the field of HRD, since the main gist of our training programs is closer to professional training in that it aims at acquiring skills and tools rather than content knowledge and education. Hence, although there are several popular models in the field of education such as the CIPP model (Asfaroh, Rosana, & Supahar, 2017; Lippe & Carter, 2018; Stufflebeam, 2003), the HRD based model is better suited for evaluating training programs. One of the most popular models for evaluation training effectiveness in HDR is the Kirkpatrick four-level model for assessment (Bates, 2004; Kirkpatrick, 1967; Praslova, 2010).

The Kirkpatrick model is considered a summative assessment model, with 4 levels of criteria: Reaction, Learning, Behavior, and Results, which are measured at the end of a training program. The Reaction level refers to how participants react to the training; e.g., their attitude towards the training program. The Learning level uses measurable indicators to determine whether learning has taken place. Behavior evaluates the implementation of the skills and knowledge acquired over the course of training in day-to-day practice. Results looks at the impact of the program in terms of goals and objectives obtained (Kirkpatrick, 1967).

Although the field of training research has grown considerably, this early model of training evaluation has maintained its popularity. Bates later criticized the four-level model (Bates, 2004; Holton, 1996) for its two embedded assumptions of a hierarchical order between the levels and causality. Bates (2004) also noted that the model failed to address both summative and formative questions, since it is administrated retrospectively.

Based on this criticism, and in order to adapt this model from management to science communication, we suggest some adaptations that address the needs of science communication training effectiveness. Rather than treating each level as constructing more useful information than the previous one, we regard each level as a "stand-alone" that can and should inform the others. We also suggest incorporating a pre-training evaluation alongside the post training evaluation, which responds to the criticism of the model's assumptions of causality and hierarchy as well as its inability to provide both a summative and formative evaluation (Alliger & Janak, 1989; Bates, 2004).

Reaction

Reaction involves capturing the participants' views, attitudes and perceptions of the training program and practice in general. In terms of science communication, reactions can be evaluated by tapping scientists' experience, attitudes, etc. towards science communication training as was done in previous studies (e.g., Besley et al., 2015). The information to address at this level is mostly based on surveys and interviews to establish attitudes or reactions after training as a summative approach, though we suggest incorporating a pre-program diagnostic assessment to identify participants' pre-course perceptions.

Learning

Learning refers to the knowledge participants acquired during the training program. In our context, this level can be used to measure participants' skills before and after the training program, or to determine the extent of the dissemination of skills and information they received during training. For example, skills such as the ability to deal with the heterogeneity of public audiences and the need to adapt their messages can be assessed. Evaluations at this level can implement self-report tools and external evaluations such as presenting a pre- and post- assignment to a target audience (via Mturk or other platforms) to evaluate, based on a validated evaluation rubric.

Behavior

Behavior refers to a change in "day-to-day" practice after training and the implementation of the knowledge addressed in the previous levels. In science communication, behavior can be evaluated by tracking participants over time to detect changes in their science communication activities, for example. A longitudinal study can also provide a great deal of information on the effectiveness of training, since behavioral change may not manifest immediately. Here again a self-report survey or interview before and after the training program can be used. In addition, peers' and experts' reviews and feedback on each participant can be gathered before or after training, or at specific time points. Changes in behavior can be observed in the actual implementation of the skills learned in the training program, but also in the participants' sense of efficacy and their willingness and actual participation in outreach and interactions with the public.

Results

The results level examines the entirety of the outcomes and evaluates them in light of the program's initial aims and goals. For example, if the training goal is to equip participants with journalistic tools, the results level will evaluate the extent to which this goal was achieved based on participants' knowledge as well as their changes in behavior, in terms of quality and quantity.

Table 8.1 maps a suggestion for integrating Baram-Tsabari and Lewenstein's (2017b) learning goals into the four-level model to evaluate a science communication program. Three of the four levels correlate with these learning goals and

Table 8.1 Mapping suggested learning goals into the evaluation model of training programs

Kirkpatrick's four level model (Kirkpatrick 1967)	*Baram-Tsabari & Lewenstein's (2017b) learning goals*	*Baram-Tsabari & Lewenstein's learning goals descriptions matching Kirkpatrick's four level model*
Reaction	Affective + Reflective + Identity goals	"Experiences excitement, interest, and motivation about science communication activities and develops attitudes supportive of effective science communication" and also "Can reflect on science and science communication's role within society [...] and on their own process of learning about and doing science communication" but also "develops an identity as someone who is able to contribute to science communication"
Learning	Content goal	"Comes to generate, understand, remember, and use concepts, explanations, arguments, models, and facts related to science communication"
Behavior	Methods + Participatory goals	"Uses science communication methods, including written, oral, and visual communication skills and tools, for fostering fruitful dialogues with diverse audiences" "Participates in scientific communication activities in authentic settings, creating written, oral and visual science messages suitable for various non-technical audiences and engaging in fruitful dialogues with those audiences"
Results	n/a	n/a

can be used as a framework in evaluating a training program that aspires to achieve these goals even partially.

Evaluation challenges

Several challenges should be taken into consideration when planning evaluations. First, science communication training programs differ in various key ways such as length, audience, venue and skills taught. In view of this variance a standardized "one-size-fits-all" evaluation is inapplicable. Each program needs to be evaluated in terms of its specific goals, target audience, skills and venue. Although positive, one of the consequences of this diversity is that generalization of the findings and evaluation of science communication training programs is problematic. Second, to date, there is no generalized "gold standard" or best practice for science communication training. We do not claim that one should exist. We do, however, argue that science communication training programs should formulate their goals and aims, and incorporate evaluations from the outset to be able to rigorously interpret the outcomes and effects of the program since "… a lack of assessment data can sometimes lead to policies and practices based on intuition, prejudice, preconceived notions, or personal proclivities – none of them desirable bases for making decisions." (Upcraft & Schuh, 2002, p. 20).

Finally, the effect of training programs takes time to "sink in" and for participants to implement what they learned. If participants' post-training skills are evaluated too close to the training program, differences may not be observed (Kraiger et al., 1993). Similarly, if we wait too long to conduct an evaluation, intervening variables can overshadow the impact of training. The question of timing is also affected by our goals in that if what interests us is change over time, assessment should be conducted at several predefined points in time. In contrast, if we want to know what the key ideas were, or the most efficient tools the participants took from the training, impact should be evaluated at the end of the program.

Key points in this chapter:

1 Science communication training programs should be accompanied by evaluations of their effectiveness and outcomes. For systematic evaluation to take place, there is a need for defined goals and measurable categories that derive from these goals. Evaluation is context-dependent and as such is based on the values that are regarded as most important by evaluators.

2 A question and answer check list for developing an evaluation. Question: "What is the most important skill/knowledge a participant in my training program should achieve by its end?". Response: "Communicate her/his research better." Then ask, "how can that be measured?" and suggest ways that can evaluate participants' ability to do so. This could include aims such as frequent use of analogies (how much is frequent?) or reduced use of jargon (use less than 2% of words unknown to the audience). After

determining the aims, a rubric adapted to the goals of the program can be constructed.

3 Using the 4-level model to construct an evaluation program for training can help map participants' changes in attitudes, skills, behavior and overall results from the program.
4 Assuming we have defined clear learning goals and measurable operational categories, what evaluation instrument will help us answer these questions? Is it appropriate for the specific training program (self vs. external reports, the skills acquired during the program, etc.)?
5 Be aware of the limitations of evaluations such as timing, funds, trained manpower, short attention span of the participants, and phrasing of questions for the survey or interview.
6 Have realistic expectations about an evaluation since the limitations can derive from the aims and goals but also relate to time and money.
7 We do not need to "re-invent the wheel" and can draw on theory and practice from other fields such as education, business management, human resources and psychology.

Notes

1 Evaluation is conducted to determine the extent to which goals are attained and their value, whereas assessment identifies the level of performance of an individual. In this context, assessment would be used to answer questions such as "what is the situation?" or "what do people know?" and evaluation answers question such as "is this situation desirable?". Since we argue here for the importance of setting goals and focus on the program level, we refer to this process as evaluation. Hence, in this chapter we use the term "evaluation" for the general (mostly summative) picture of science communication training programs and "assessment" for specific methods that feed into the process of evaluation.

2 1. "Share the findings and excitement of science" 2. "Increase appreciation for science as a useful way of understanding and navigating the modern world...." 3. "... Increase knowledge and understanding of science related to a specific issue" that requires a decision, 4. "... Influence people's opinions, behavior, and policy preferences... when the weight of evidence clearly shows that some choices (...) have consequences for public health, public safety, or some other societal concern", and 5. "... Engage with diverse groups so their perspectives about science" (particularly on contentious issues) "can be considered in seeking solutions to societal problems that affect everyone" (National Academies of Sciences Engineering and Medicine, 2017).

3 An evaluation sheet can contain practical categories such as the use of hand gestures to demonstrate or elucidate an explanation during the presentation, or conceptual categories such as building a story arc for a presentation. The categories and evaluation rubric are defined by the goals and agendas of the training program.

References

Alliger, G. M., & Janak, E. A. (1989). Kirkpatrick's Levels of Training Criteria: Thirty Years Later. *Personnel Psychology*, *42*(2), 331–342.

Asfaroh, J. A., Rosana, D., & Supahar. (2017). Development of CIPP Model of Evaluation Instrument on the Implementation of Project Assessment in Science Learning. *International Journal of Environmental and Science Education*, *12*(9), 1999–2010.

Atkins, P. W. B., & Wood, R. E. (2002). Self Versus Others' Ratings As Predictors Of Assessment Center Ratings: Validation Evidence For 360-Degree Feedback Programs. *Personnel Psychology*, *55*(4), 871–904.

Baldwin, W. (1999). Information No One Else Knows: The Value of Self-Report. In A. A. Stone, C. A. Cachrach, J. B. Jobe, H. S. Kurtzman & V. S. Cain (Eds.), *The Science of Self-report: Implications for Research and Practice* (pp. 3–8). USA: Lawrence Erlbum Associates, Inc., 3–8.

Baram-Tsabari, A., & Lewenstein, B. V. (2012). An Instrument for Assessing Scientists' Written Skills in Public Communication of Science. *Science Communication.*

Baram-Tsabari, A., & Lewenstein, B. V. (2016). Assessment. In S. Mvd & M. J. DV, (Eds.), *Science and Technology Education and Communication: Seeking Synergy* (pp 161–184). Rotterdam: Sense Publishers.

Baram-Tsabari, A., & Lewenstein, B. V. (2017a). Preparing Scientists to Be Science Communicators. In P. G. Patrick (Ed.), *Preparing Informal Science Educators* (pp. 437–471). Cham: Springer International Publishing.

Baram-Tsabari, A., & Lewenstein, B. V. (2017b). Science Communication Training: What are We Trying to Teach? *International Journal of Science Education, Part B*, 1–16.

Baruch, Y., & Holtom, B. C. (2008). Survey response rate levels and trends in organizational research. *Human Relations*, 61 (8), 1139–1160.

Bates, R. (2004). A critical analysis of evaluation practice: the Kirkpatrick model and the principle of beneficence. *Evaluation and Program Planning*, *27*(3), 341–347.

Besley, J. C. (2015a). Predictors of Perceptions of Scientists: Comparing 2001 and 2012. *Bulletin of Science, Technology & Society*, *35*(1–2), 3–15.

Besley, J. C. (2015b). What do scientists think about the public and does it matter to their online engagement? *Science and Public Policy*, *42*(2), 201–214.

Besley, J. C., Dudo, A., & Storksdieck, M. (2015). Scientists' views about communication training. *Journal of Research in Science Teaching*, *52*(2), 199–220.

Besley, J. C., Dudo, A., Yuan, S., & Lawrence, F. (2018). Understanding Scientists' Willingness to Engage. *Science Communication*, 107554701878656.

Besley, J. C., Dudo, A. D., Yuan, S., & Abi Ghannam, N. (2016). Qualitative Interviews With Science Communication Trainers About Communication Objectives and Goals. *Science Communication*, *38*(3), 356–381.

Besley, J. C., Oh, S. H., & Nisbet, M. (2013). Predicting scientists' participation in public life. *Public understanding of science (Bristol, England)*, *22*(8), 971–987.

Besley, J. C. & Nisbet, M. (2013). How scientists view the public, the media and the political process. *Public Understanding of Science*, *22*(6), 644–659.

Besley, J. C., & Tanner, A. H. (2011). What Science Communication Scholars Think About Training Scientists to Communicate. *Science Communication*, *33*(2), 239–263.

Black, P., & Wiliam, D. (1998). Assessment and Classroom Learning. *Assessment in Education: Principles, Policy & Practice*, *5*(1), 7–74.

Black, P. J. (1993). Formative and Summative Assessment by Teachers. *Studies in Science Education*, *21*(1), 49–97.

Bogue, B., Shanahan, B., Marra, R. M., & Cady, E. T. (2013). Outcomes-Based Assessment: Driving Outreach Program Effectiveness. *Leadership and Management in Engineering*, *13*(1), 27–34.

Bracken, D. W., Timmreck, C. W., Fleenor, J. W., & Summers, L. (2001). 360 Feedback from Another Angle. *Human Resource Management*, *40*(1), 3–20.

Brownell, S. E., Price, J. V, & Steinman, L. (2013). A writing-intensive course improves biology undergraduates' perception and confidence of their abilities to read scientific literature and communicate science. *Advances in Physiology Education, 37*(1), 70–79.

Brutus, S., Aguinis, H., & Wassmer, U. (2013). Self-Reported Limitations and Future Directions in Scholarly Reports. *Journal of Management, 39*(1), 48–75.

Burden, S., Topping, A., & O'Halloran, C. (2015). The value of artefacts in stimulated-recall interviews. *Nurse Researcher, 23*(1), 26–33.

Cameron, C., Collie, C. L., Baldwin, C. D., Bartholomew, L. K., Palmer, J. L., Greer, M., & Chang, S. (2013). The development of scientific communication skills: a qualitative study of the perceptions of trainees and their mentors. *Academic medicine : journal of the Association of American Medical Colleges, 88*(10), 1499–1506.

Cerrato, S., Daelli, V., Pertot, H., Puccioni, O. (2018). The public-engaged scientists: Motivations, enablers and barriers. *Research for All, 2*(2), 313–322.

Chapelle, C.A., Cotos, E., & Lee, J. (2015). Validity arguments for diagnostic assessment using automated writing evaluation. *Language Testing, 32*(3), 385–405.

Creswell, J. (2003). *Research design: Qualitative, quantitative, and mixed methods approaches*. Thousand Oaks, CA: Sage.

Dudo, A. (2012). Toward a Model of Scientists' Public Communication Activity: The Case of Biomedical Researchers. *Science Communication, 35*(4), 476–501.

Dudo, A., & Besley, J. J. C. (2016). Scientists' Prioritization of Communication Objectives for Public Engagement. *PLoS One, 11*(2), e0148867.

Dunwoody, S., Brossard, D., & Dudo, A. (2009). Socialization or Rewards? Predicting U.S. Scientist-Media Interactions. *Journalism & Mass Communication Quarterly, 86*(2), 299–314.

Ecklund, E. H., James, S. A., and Lincoln, A. E. (2012). How academic biologists and physicists view science outreach. *PloS One, 7*(5), e36240.

Eliot & Associates. (2007). Guidelines for Conducting a Focus Group.

Entradas, M. & Bauer, M. M. (2016). Mobilisation for public engagement: Benchmarking the practices of research institutes. *Public understanding of science*, doi: 10.1177/0963662516633834

Grand, A., Davies, G., Holliman, R., & Adams, A. (2015). Mapping public engagement with research in a UK University. *PloS One, 10*(4), e0121874.

Haran, B., & Poliakof, M. (2011). How to measure the impact of chemistry on the small screen. *Nature Chemistry, 3*, 180–182.

Harlen, W., & James, M. (1997). Assessment and Learning: differences and relationships between formative and summative assessment. *Assessment in Education: Principles, Policy & Practice, 4*(3), 365–379.

Holt, R. W., Boehm-Davis, D. A., & Beaubien, J. M. (2001). Evaluating Resource Management Training. In E. Salas, C. A. Bowers, & E. Edens, (Eds.), *Improving Teamwork in Organizations: Applications of Resource Management Training* (pp. 165–190). Boca Raton: CRC Press.

Holton, E. F. (1996). The flawed four-level evaluation model. *Human Resource Development Quarterly, 7*(1), 5–21.

Jonsson, A., & Svingby, G. (2007). The use of scoring rubrics: Reliability, validity and educational consequences. *Educational Research Review, 2*(2), 130–144.

Khandker, S. R. (2010). *Handbook on Impact Evaluation: Quantitative Methods and Practices*. 1st edition. Washington, DC: World Bank.

Kirkpatrick, D. L. (1967). Evaluation of training. In R. L. Craig & L. R. Bittel (Eds.), *Training and Development Handbook* (pp. 87–112). New York: McGraw Hill.

Kraiger, K., Ford, J. K., & Salas, E. (1993). Application of Cognitive, Skill-Based, and Affective Theories of Learning Outcomes to New Methods of Training Evaluation. *Journal of Applied Psychology*, *78*(2), 311–328.

Lippe, M., & Carter, P. (2018). Using the CIPP Model to Assess Nursing Education Program Quality and Merit. *Teaching and Learning in Nursing*, *13*(1), 9–13.

Lyle, J. (2003). Stimulated recall: a report on its use in naturalistic research. *British Educational Research Journal*, *29*(6), 861–878.

Masadeh, M. A. (2012). Focus Group: Reviews and Practices. *International Journal of Applied Science and Technology*, *2*(10).

McCann, B. M., Cramer, C. B., & Taylor, L. G. (2015). Assessing the Impact of Education and Outreach Activities on Research Scientists. *Journal of Higher Education Outreach and Engagement*, *19*(1), 65–78.

McGillin, V. (2003). Research versus Assessment: What's the Difference? [online]. *Academic Advising Today*. Available from: www.nacada.ksu.edu/Resources/Academic-Advising-Today/View-Articles/Research-versus-Assessment-Whats-the-Difference.aspx [Accessed 22 Jul 2018].

Mercer-Mapstone, L., & Kuchel, L. (2015a). Teaching Scientists to Communicate: Evidence-based assessment for undergraduate science education. *International Journal of Science Education*, *37*(10), 1613–1638.

Mercer-Mapstone, L., & Kuchel, L. (2015b). Core Skills for Effective Science Communication: A Teaching Resource for Undergraduate Science Education. *International Journal of Science Education, Part B*, 1–21.

Metcalfe, J. E. & Gascoigne, T. (2009). Teaching Scientists to Interact with the Media. *Issues*, *87*.

Montgomery, S. L. (2017). *The Chicago Guide to Communicating Science*. 2nd Edition. Chicago: University of Chicago Press.

Nagle, B., & Williams, N. (2004). Methodology Brief: Introduction to Focus Groups. *Extension Community and Economic Development Publications*, *7*.

National Academies of Sciences Engineering and Medicine. (2017). *Communicating Science Effectively: A Research Agenda*.

Nisbet, M. C., & Scheufele, D.A. (2009). What's next for science communication? *American Journal of Botany*, *96*(10), 1767–1778.

Patton, M. Q. (2015). *Qualitative research & evaluation methods: integrating theory and practice*. Saint Paul, MN: Sage.

Peterman, K., Robertson Evia, J., Cloyd, E., & Besley, J. C. (2017). Assessing Public Engagement Outcomes by the Use of an Outcome Expectations Scale for Scientists. *Science Communication*, *39*(6), 782–797.

Peters, H.P. (2012). Gap between science and media revisited: Scientists as public communicators. *Proceedings of the National Academy of Sciences of the United States of America*, *110* (Supplement_3), 14102–14109.

Poliakoff, E., & Webb, T. L. (2007). What Factors Predict Scientists' Intentions to Participate in Public Engagement of Science Activities? *Science Communication*, *29*(2), 242–263.

Praslova, L. (2010). Adaptation of Kirkpatrick's four level model of training criteria to assessment of learning outcomes and program evaluation in Higher Education. *Educational Assessment, Evaluation and Accountability*, *22*(3), 215–225.

Raizen, S. A., & Rossi, P. H. (1981). *Program evaluation in education, when? how? to what ends?* Washington, DC 20418: National Academy Press.

Rakedzon, T. and Baram-Tsabari, A. (2017). Assessing and improving L2 graduate students' popular science and academic writing in an academic writing course. *Educational Psychology, 37*(1), 48–66.

Rakedzon, T., Segev, E., Chapnik, N., Yosef, R., & Baram-Tsabari, A. (2017). Automatic jargon identifier for scientists engaging with the public and science communication educators. *PLoS One, 12*(8), e0181742.

Robertson Evia, J., Peterman, K., Cloyd, E., & Besley, J. (2018). Validating a scale that measures scientists' self-efficacy for public engagement with science. *International Journal of Science Education, Part B, 8*(1), 40–52.

Rodgers, S., Wang, Z., Maras, M. A., Burgoyne, S., Balakrishnan, B., Stemmle, J., & Schultz, J. C. (2018). Decoding Science: Development and Evaluation of a Science Communication Training Program Using a Triangulated Framework. *Science Communication, 40*(1), 3–32.

Rowland, S., Hardy, J., Colthorpe, K., Pedwell, R., & Kuchel, L. (2018). CLIPS (Communication Learning in Practice for Scientists): A New Online Resource Leverages Assessment to Help Students and Academics Improve Science Communication. *Journal of Microbiology & Biology Education, 19*(1).

Royal Society, The. (2006). *Survey of factors affecting science communication by scientists and engineers*. London: The Royal Society

Salas, E., Tannenbaum, S. I., Kraiger, K., & Smith-Jentsch, K. A. (2012). The Science of Training and Development in Organizations: What Matters in Practice. *Psychological Science in the Public Interest, 13*(2), 74–101.

Sevian, H., & Gonsalves, L. (2008). Analysing how Scientists Explain their Research: A rubric for measuring the effectiveness of scientific explanations. *International Journal of Science Education, 30*(11), 1441–1467.

Sheehan, K. B. (2006). E-mail Survey Response Rates: A Review. *Journal of Computer-Mediated Communication, 6*(2), 0–0.

Shubert, C. W., & Meredith, D. C. (2015). Stimulated recall interviews for describing pragmatic epistemology. *Physical Review Special Topics – Physics Education Research, 11*(2), 020138.

Stone, A. A., Turkkan, J., Bacharach, C. A., Jobe, J. B., Kurtzman, H. S., & Cain, V. S. (1999). *The Science of Self Report: Implications for Research and Practice*. Lawrence Erlbaum Associates, Inc.

Stufflebeam, D. L. (2003). The CIPP Model for Evaluation. In T. Kellaghan & D. L. Stufflebeam (Eds.), *The International Handbook of Educational Evaluation* (pp. 31–62). Dordrecht: Springer Netherlands.

Tucker, J. M. (2014). Stop Asking Students to "Strongly Agree": Let's Directly Measure Cocurricular Learning. *About Campus, 19*(4), 29–32.

Upcraft, M. L., & Schuh, J. H. (2002). Assessment Vs. Research why we Should Care about the Difference. https://doi.org/10.1177/108648220200700104.

Wiliam, D., & Black, P. (1996). Meanings and Consequences: a basis for distinguishing formative and summative functions of assessment? *British Educational Research Journal, 22*(5), 537–548.

Part III

Future directions for science communication training

9 Abandoning the runaway train

Slowing down to draw on lessons learned from health communication training

Brenda L. MacArthur, Nicole J. Leavey, and Amanda E. Ng

The scientific community is experiencing threats to government funding (Mervis, 2017), the general public is basing decisions on scientific myths and misinformation (Vosoughi, Roy & Aral, 2018), and on top of that, many scientists mistakenly believe that the public no longer trusts science (Jackson, 2018). Scientists have not prioritized communication as a part of the scientific process, and as a result the divide between the scientific community and the public continues to grow (Rajput, 2017). As the fate of the field is being called into question, scientists are ready to acknowledge the severity of the problem, and recognize that drastic measures are needed to remedy growing concerns. Exponential growth in the science communication training field reflects this shift, as scientists are increasingly seeking out opportunities to capitalize on learning effective communication practices. As the demand for science communication training continues to rise, it is of utmost importance that trainings are designed to address the core issues facing the scientific community. Research has explored the development of such programs and the response from the scientific community (Besley, Dudo, & Storksdieck, 2015; Besley, Dudo, Yuan, & Ghannam, 2016; Trench & Miller, 2012). Such rapid growth has resulted in a highly diverse expanse of training programs (Baram-Tsabari & Lewenstein, 2017). These trainings embody both traditional didactic approaches and more experiential learning environments (Yuan et al., 2017). It was not long ago that the health communication field exploded in a similar manner, prompting health care professionals to seek out communication training to foster enhanced interactions with patients.

Drawing on health communication

The parallels between the science and health communication fields make it nearly impossible to avoid drawing comparisons, especially when there are lessons to be gleaned. First, the health communication field emerged as a result of complaints from dissatisfied patients concerning their quality of care. Today, the science communication field too is growing in response to threats to the field. Second, traditional approaches to health care required patients to heed provider instructions without any explanation, a process that fueled non-adherence among patients and loss of trust in health care professionals (Kerse et al., 2004). Similarly, scientists often expect the public to adhere to behavioral and policy recommendations, but do not offer clear and concise explanations to help them better understand and assign deeper, personal meaning to those findings. Third, both fields experienced a shift away from traditionally held beliefs about communication being a "soft" skill compared to technical knowledge and abilities. Given the health communication field's success and sustained growth, the science communication training community has a unique opportunity to capitalize on best practices and avoid pitfalls uncovered in health communication training.

This chapter will explore three key lessons that science communication trainers can learn from the health communication field and offer key takeaways that can implemented for immediate action.

Lesson 1: accepting the burden of proof rather than assigning blame

Science is complex, as a result people will use cognitive shortcuts or heuristic approaches to understand information (Akin & Scheufele, 2017; Scheufele, 2006;) – this often leads the public to ask more questions. Communication too is complex, and at times, those questions can function to challenge a particular research project, the scientists conducting that research, and/or the field in general. When this occurs, scientists typically assign blame to the members of the public. The lay person is blamed for a lack of knowledge about science (Davies, 2008), the media are blamed for misrepresenting scientific information, and elected officials are accused of promoting extremist views (Pew Research Center, 2017). In all of these instances, the burden of proof does not fall to the scientific community to help these audiences truly understand and assign personal meaning to scientific findings.

Scientists' perceptions of the public are not unique to the field. Health care professionals too have been known to place blame on non-expert audiences to explain or justify confusion about complex health information. Patients who do not adhere to physician recommendations are often labeled as "non-compliant," while those who pose too many questions are "difficult" (Russell, Daly, Hughes & Op't Hoog, 2003). Caregivers who want additional clarification are often noted as being overbearing (Laidsaar-Powell, Butow, Boyle & Juraskova, 2018), and online health sources (aka "Dr. Google") distribute inaccurate health information (Benigeri & Pluye, 2003; Fisher, O'Connor, Flexman, Shapera & Ryerson, 2016). In health care, the demand for communication training emerged in part, as a way for healthcare professionals to take more responsibility for communicating health information that non-expert audiences could understand. Much of health communication training is focused on patient-centered communication, which requires health care professionals to consider patients' physical symptoms (i.e., the scientific information), in concert with their personal experiences of those symptoms (i.e., emotions, concerns, feelings) (Dean & Street,

2016). In this way, health care professionals understand the patient as a "whole person" and are better prepared to communicate complex health information in ways that are most likely to resonate with that patient. Through patient-centered communication, health care professionals can uncover common ground, build trust to enhance professional relationships with patients, and employ targeted messaging to overcome communication barriers unique to the patient (Stewart et al., 1995).

It is important for the scientific community to draw on these lessons. Science communication training must teach scientists how to take an audience-centered approach to sharing complex scientific information with non-expert audiences. Similar to patient-centered communication, this approach requires scientists to place emphasis on understanding their audiences' interests, concerns, and perceptions of science, and use that information to frame scientific information in a way that is most relevant and personal to their audience(s) (Rogers, 2000). This approach requires a shift away from the deficit model of thinking, which suggests that gaps between scientists and the public are a result of lack of information or knowledge (Layton, Jenkins, McGill, & Davey, 1993; Wynne, 1991; Ziman, 1991). The audience-centered approach to science communication proposes a model where the burden of proof rests with scientists, to engage in more meaningful interactions with various audiences rather than simply fill the knowledge deficit with more information (Bauer, Allum, & Miller, 2007). Humans are complex beings. Our personal experiences and cultural, economic, and political worldviews all shape our attitudes toward science (Sturgin & Allum, 2004), so it is of utmost importance that scientists consider these influences to be effective communicators.

What can the scientific community do to pivot away from the deficit model thinking toward a more holistic approach to science communication? Let's explore that in Lesson 2.

Lesson 2: engaging key audiences for audience-centered training

In science communication we typically think of the *audience* as those individuals consuming scientific information. In the training context, the scientists become the audience consuming the training content. In the process of designing and delivering evidence-based training, the trainers must become the audience to the scientific community. At one point or another we all become *the audience*. The literature consistently supports the integration of trainees' perspectives into the communication training development process to ensure a strong evidence-base (Brown & Bylund, 2008; Kreps, 2014; Neuhauser & Paul, 2011). Yet the integration of other audiences' perspectives receives considerably less attention. With concerns mounting over the state of the field, science communication training has grown exponentially, resulting in an expanse of training offerings (Baram-Tsabari & Lewenstein, 2017). To accommodate this demand, many science communication trainings approach scientists with the assumption that there are already long-term goals identified; as a result the training efforts focus on specific communication skills (Besley et al., 2016), but in doing so may be missing a key piece of the puzzle.

The health communication field experienced similar increases in demand for communication training when health care professionals realized that care quality had decreased and financial consequences ensued. Communication training was mobilized in a similar fashion, as a quick remedy to concerns related to care quality. Much of this training focused primarily on the ability to simplify language to educate patients about complex health information (Sudore & Schillinger, 2009), and elicited small effect sizes that were not sustained over time (Kreps, 2014). It was not until patient-centered communication was introduced that trainers realized effective communication was much more than just designing a simple message. For patients to be motivated to act on that information, they needed not just to understand the information, but also to assign personal meaning and relevance to it (Stubblefield, 1997; Hall & Johnson-Turbes, 2015). For them, quality of care ratings was not just about what was said, but more about the circumstances under which that information was communicated (Donabedian, 1992). There are two important lessons to be learned here. First, teaching scientists how to simplify complex language may improve the public's understanding of a scientific concept in the short term. But in the absence of acknowledging its relevance and assigning personal meaning to it, that understanding is not likely to elicit any motivation for continued engagement or action (Cook & Artino, 2016). Second, in designing evidence-based trainings, trainers need to integrate the target audience's perspective with social science research. Such processes will ensure that trainings are actually achieving and sustaining the long-term goals they are targeting.

Training goals

Similar to health care professionals, scientists find great value in trainings that target micro-level communication competencies such as message comprehension and source credibility (Besley et al., 2015, 2016). Mastering these communication competencies can help scientists achieve traditional goals to disseminate information and fill the knowledge deficit in the short-term. Yet they do not speak to longer-term goals of the field, such as increasing public engagement for more meaningful interactions between the scientific community and the general public. The National Academy of Sciences' Sackler colloquiums on the "science of science communication" connect scientists to the social science behind science communication (Besley et al., 2015), much like the field of health communication connected health care professionals to the social science behind medical interactions. This research, like the health communication literature, consistently highlights the value of pivoting toward audience-centered communication goals. It explains that scientists who *engage* with the public leverage communication to foster authentic interactions, build interest and excitement around science, motivate their audience(s) to act, and establish trust in science

(Storksdieck, Stein, & Dancu, 2006). The National Academy of Science identifies five goals for communicating science (National Academies of Sciences, Engineering, and Medicine, 2017, p. 2):

1 Simply to share the finding and excitement of science.
2 To increase appreciation for science as a useful way of understanding and navigating the modern world.
3 To increase knowledge and understanding of the science related to a specific issue.
4 To influence people's opinions, behavior, and policy preferences.
5 To engage with diverse groups so that their perspectives about science related to important social issues can be considered in seeking solutions to societal problems that affect everyone.

Unfortunately, the research also suggests that many scientists tend to hold traditional views about the value of science communication training. They do not place high value on learning skills that will help them to be viewed as responsive, caring, and concerned. They perceive the ability to reframe science in ways that resonate with their audience(s) as the least valuable, least ethical, and least achievable communication skill (Besley et al., 2015). Moreover, in a recent study, scientists identified "defending science" and "informing the public about science" as their top goals in seeking out communication training, while "exciting the public about science," "building trust," and "tailoring messages" were less highly ranked (Dudo & Besley, 2016). These findings suggest a clear disconnect between what scientists and social science researchers believe communication training can offer the field.

Integrating different perspectives

Scientists prefer that communication trainings teach skills that offer a quick remedy to the problem – to fill the knowledge deficit. Science communication scholars suggest teaching skills that can be applied to broader audience-centered goals such as public engagement, which are more likely to be sustained over time. The latter is a much more complicated process, which requires increased effort, prolonged exposure, and integration of communication throughout the scientific process (Bankston & McDowell, 2018). We know that before scientists will buy into communication training targeting public engagement, they must first accept that it has the power to remedy their growing concerns about the field (Besley et al., 2015). Thus, one of the primary goals in science communication training programs should be to address the value of public engagement. Scientists need to understand how this longer-term goal and shift in traditional way of thinking about communication, has the power to remedy their growing concerns about the field. If trainers could frame science communication as audience-centered verses information transfer, scientists may be more likely to recognize the value of public engagement and invest in the process. Research suggests that scientists' external efficacy (i.e., their belief that public engagement can make a difference, combined with the perceived ethicality of specific training goals), is the best predictor of the value they place on such goal-oriented training (Besley et al., 2015).

In the case of health communication, it was not until health care professionals moved away from the longstanding paternalistic model of care toward a more patient-centered approach that communication became widely accepted and integrated as a part of the health care process. Scientists are in a similar position to abandon the traditional deficit model of thinking and move toward audience-centered approaches that integrate public engagement into the scientific process. If scientists are leveraging communication training to address the core issues facing the scientific community, they must adopt this audience-centered approach to communication for any changes to be sustained in the long run. As we learned from the health communication field, audience-centered trainings that take a holistic approach to communication have the power to change the status quo and sustain that change over time.

Lesson 3: capitalizing on existing experiential learning: a path to implementation

Scientists are much more likely to retain information that they can experience versus information they passively consume. Experiential science communication training targets skills that drive critical thinking and problem solving, rather than didactic approaches that rely on mechanical learning and memorization (Gorghiu & Santi, 2016). Through practicing newly acquired communication skills, scientists assign a deeper mental meaning to the links they personally observe between abstract concepts and practical application leading to sustained behavior change (Dewey, 1938; Kolb, 1984). There is a general consensus about the benefits that science communication training brings to the scientific community (Besley & Tanner, 2011), yet logistical challenges often make the implementation of such trainings, especially those that are experiential in nature, less frequent. The problem that science communication trainers face today, which is largely the same problem health communication trainers experienced, is the limited time that scientists can devote to communication training, especially in a system that is not set up to reward participation. In this final section we will highlight one of the most important lessons the health communication training field has to offer – the integration of communication training into the structure of medical education and health care delivery.

Formal education

The National Academy of Medicine identifies the ability to deliver patient-centered care as a core competency for all health care professionals (Makoul, 2001; National Academy of Medicine, 2018). This distinction fueled the integration of health communication into the health sciences, and today communication courses are increasingly being included as part of core requirements in medical education (Bennett & Lyons, 2011; Cegala & Broz, 2002; Norgaard, Ammentorp, Kyvik, & Kofoed, 2012). That said, many of these courses were introduced in didactic formats, a challenge that many educators still struggle with today. Health communication researchers have always stressed that because communication skills need to be practiced over time to be mastered, training programs should include as much realism as possible to be highly effective (Kreps & Thornton, 1992; Ross, 2012). The problem is that experiential learning takes time and repeated exposure to be highly successful (Koponen, Pyorala, & Isotalus, 2012). As a result, didactic training models became ideal alternatives in medical education, because of their ease of use, cost effectiveness, and one-to-many designs that can be offered via mediated platforms (Roter et al., 2012). But health communication researchers strongly caution that although such trainings can successfully teach cognitive skills, face-to-face instruction and practice is needed to acquire and sustain affective communication skills (Wittenberg-Lyles, Goldsmith, Ferrell, & Burchett, 2014). While health communication trainers still struggle with moving beyond traditional approaches to teaching cognitive skills in medical education, they have found ways to pair these approaches with existing opportunities for experiential learning.

Leverage existing opportunities

In science education, credit-heavy requirements make communication courses difficult to add into the existing structure (Train & Miyamoto, 2017). One way to approach this challenge is to integrate science communication skills into existing courses that teach scientific content (Brownell, Price, & Steinman, 2013). Doing so would highlight communication as an integrated element of the scientific process rather than an add-on to a program of study, or worse, an afterthought once the program is completed. Similar integration took place in medical education before the National Academy of Medicine identified communication as a core competency for the field. Health communication trainers were also strategic about capitalizing on existing structures to ensure that communication training included experiential learning opportunities.

Similar to flight simulation training for airline pilots, clinical simulation centers simulate actual patient interactions by utilizing actors to portray patients. Trainees practice delivering care to "standardized patients" and are evaluated on medical knowledge and technical abilities. Today most simulation centers also evaluate trainees on patient-centered communication competencies (Arnold, McKenzie, Miller & Mancini, 2018). After completing this type of scenario-based practice, health care professionals have demonstrated improved communication in clinical practice (Hsu, Chang, & Hsieh, 2015), and reported increased self-efficacy (Norgaard et al., 2012) and confidence (Brown et al., 2010; Hahn & Cadogan, 2011). For health care professionals, didactic communication training may be ideal for learning cognitive skills but the follow-up requirement to then practice or "experience" such skills in clinical simulation centers solidifies the learning experience. Health care professionals who practice communication skills are three times more likely to retain them, simply because their personal experience evolves cognitive skills into insight (Gaffney, Farnan, Hirsch, McGinty, & Arora, 2016). Science communication trainers could borrow this practice to capitalize on existing experiential learning requirements in science education. Although these requirements vary by specialty, many science degrees require some sort of field training, and many social science degrees require an outreach component. Bringing scientists and social scientists together to fulfill experiential requirements could provide a unique opportunity for high quality science communication training, with the added benefit of exposure to transdisciplinary collaboration. Such a training scenario would allow for science communication skills to be taught in the context where they will eventually be applied, a practice that is cited as most effective in eliciting behavior change through communication training (Craig et al., 2008; Lipsey, 1993; Sidani & Sechrest, 1999).

Another way health communication trainers leverage existing structures for experiential learning in medicine is through interactive, panel discussions called Schwartz Rounds®. These relatively informal sessions take place during the lunch hour and bring together a variety of health care professionals, including those from the medical humanities and social sciences, sometimes even patients. Moderated discussions encourage participants to draw on their shared human experience and discuss social and emotional issues they face in patient care, an uncommon practice in their professional role (Groopman, 2007). These sessions build on the idea that patient-centered communication is more than just the ability to deliver a comprehensible health message. For it to ultimately improve patient-reported health care quality, the communication must come from a place of genuine connection that takes into account the patient as a whole person (Donabedian, 1992). It is a skill that can only be learned through first-hand experience (Kolb, 1984). For scientists, the idea of audience-centered communication is very similar. If the broad goal of science communication training is to ultimately facilitate public engagement, then scientists must be able to connect with different public audiences on a basic human level. If scientists perceive public communication as an opportunity to give back to society, as trainers suggest (Yuan et al., 2017), then they will likely support this audience-centered communication approach. Experiential learning platforms similar to Schwartz Rounds® could be the first step in this process, to not only help scientists recognize facets of themselves beyond their professional role, but also help them recognize the value of connection.

Professional development

Health communication trainers have overcome these challenges by finding ways to compensate health care professionals for their time in ways that can be applied to their professional obligations. Unlike science communication training, health communication training is routinely offered as a professional development opportunity, known as Continuing Medical Education (CME,) for health care professionals who are already in practice. Each year health care professionals must complete a minimum number of CME credits to maintain licensure and certification. These programs are approved through various accrediting bodies specific to each health profession. Through accreditation, health professionals are incentivized to participate in communication training because they can justify the time spent as a professional development activity. Although no formal professional development is required for scientists, finding ways to incentivize them through professional channels can serve a similar purpose. For example, most scientists have access to funding through their universities or organizations to support their participation in professional conferences. If science communication training was offered as a pre-conference workshop or main conference sessions, scientists might be more incentivized to participate as a part of their attendance at an annual meeting.

The design and implementation of high quality, experiential communication training places an additional burden on already limited time and financial resources in both the health and science fields. Health communication trainers have found ways to deal with these issues by leveraging existing educational structures based in experiential learning and offering training as professional development opportunities. As we've seen, although didactic approaches to communication training are not ideal, they often cannot be avoided. Integrating audience-centered communication skills into existing experiential learning opportunities reduces cost and maximizes time while maintaining a high-quality learning environment.

Conclusion

The health communication field has grown leaps and bounds in recent decades, largely because the medical system has formally acknowledged its value and identified patient-centered communication as a core competency for all health care professionals. The entire field is slowly pivoting away from traditional paternalistic approaches to medicine toward patient-centered approaches that heavily rely on communication as a key for success. The sciences have yet to experience this type of cultural shift. However, the rise of the "science of science communication" field, coupled with the support and recognition it has received from the National Academy of Sciences, suggests the potential for a similar cultural shift. This is a pivotal moment for the future of the science field. Scientists must join forces with science communication researchers, just as health care professionals did with health communication scholars, to promote audience-centered communication as the pathway to greater public understanding of science, greater public engagement with science, and ultimately greater public support for science.

References

Arnold, J. L., McKenzie, F. D., Miller, J. L., & Mancini, M. E. (2018). The many faces of patient-centered simulation. *Simulation in Healthcare: The Journal of the Society for Simulation in Healthcare, 13*, 51–55. doi: 10.1097/SIH.0000000000000312.

Akin, H., & Scheufele, D. A. (2017). Overview of the Science of Science Communication. In: T*he Oxford Handbook of The Science of Science Communication* (pp. 25–33). Oxford: Oxford University Press,.

Bankston, A., & McDowell, G. S. (2018). Changing the culture of science communication training for junior scientists. *Journal of Microbiology & Biology Education, 19*,1–6. doi: 10.1128/jmbe.v19i1.1413

Baram-Tsabari, A., & Lewenstein, B. V. (2017). Science communication training: What are we trying to teach? *International Journal of Science Education, 7*(3), 285–300. doi: 10.1080/21548455.2017.1303756

Bauer, M. W., Allum, N., & Miller, S. (2007). What can we learn from 25 years of PUS survey research? Liberating and expanding the agenda. *Public Understanding of Science, 16*, 79–95. doi: 10.1177/0963662506071287

Benigeri, M., & Pluye, P. (2003). Shortcomings of health information on the internet. *Health Promotion International, 18*, 381–386. doi: 10.1093/heapro/dag409

Bennett, K., & Lyons, Z. (2011). Communication skills in medical education: An integrated approach. *Education Research and Perspectives, 38*, 45–56.

Besley, J., & Tanner, A. (2011). What science communication scholars think about training scientists to communicate. *Science Communication, 33*, 239–263.

Besley, J., Dudo, A., & Storksdieck, M. (2015). Scientists' views about communication training. *Journal of Research in Science Teaching, 52*, 199–220.

Besley, J., Dudo, A., Yuan, S., & Ghannam, N. A. (2016). Qualitative interviews with science communication trainers about communication objectives and goals. *Science Communication, 38*, 356–381.

Brown, R. F., & Bylund, C. L. (2008). Communication skills training: Describing a new conceptual model. *Academic Medicine, 83*, 37–44.

Brown, R. F., Bylund, C. L., Gueguen, J. A., Diamond, C., Eddington, J., & Kissane, D. (2010). Developing communication skills training for oncologists: Describing the content and efficacy of training. *Communication Education, 59*, 235–248.

Brownell, S. E., Price, J. V., & Steinman, L. (2013). Science communication to the general public: Why we need to teach undergraduate and graduate students this skill as part of their formal scientific training. *Journal of Undergraduate Neuroscience Education, 12*, 6–10.

Cegala, D. J., & Broz, S. L. (2002). Physician communication skills training: A review of the theoretical backgrounds, objectives and skills. *Medical Education, 36*, 1004–1016.

Cook, D. A., & Artino, A. R. (2016). Motivation to learn: An overview of contemporary theories. *Medical Education, 50*, 997–1014. 10.1111/medu.13074

Craig, P., Dieppe, P., Macintyre, S., Michie, S., Nazareth, I., & Petticrew, M. (2008). Developing and evaluating complex interventions: the new Medical Research Council guidance. *British Medical Journal*mj, *337*, a1655.

Davies, S. R. (2008). Constructing communication: Talking to scientists about talking to the public. *Science Communication, 29*, 413–434. doi: 10.1177/1075547009316222

Dean, M., & Street, R. L. (2016). Patient-centered communication. In E. Wittenberg, B. Ferrell, J. Goldsmith, T. Smith, S. Ragan, M. Glajchen & G. Handzo (Eds.), *Textbook of Palliative Care Communication* (pp. 238–245). New York, NY: Oxford University Press.

Dewey, J. (1938). *Experience and Education.* New York: Macmillan.

Donabedian A. (1992). Quality assurance in health care: Consumers' role (The Lichfield Lecture). *Quality Health Care, 1*, 247–251.

Dudo, A., & Besley, J. C. (2016). Scientists' prioritization of communication objectives for public engagement. *PLoS ONE, 11*, 1–18.

Fisher, J. H., O'Connor, D., Flexman, A. M., Shapera, S., & Ryerson, C. J. (2016). Accuracy and reliability of internet resources for information on idiopathic pulmonary fibrosis. *American Journal of Respiratory and Critical Care Medicine, 194*. doi: 10.1164/rccm.201512-2393OC

Gaffney, S., Farnan, J. M., Hirsch, K., McGinty, M., & Arora, V. M. (2016). The modified, multi-patient observed simulated handoff experience (M-OSHE): Assessment and feedback for entering residents on handoff performance. *Journal of General Internal Medicine. 31*, 438–441.

Gorghiu, G., & Santi, E. (2016). Applications of experiential learning in science education non-formal contexts. *The European Proceedings of Social & Behavioural Sciences*, 320–326. doi: 10.15405/epsbs.2016.11.33.

Groopman, J. (2007). *How doctors think.* Boston, MA: Houghton Mifflin.

Hahn, J. E., & Cadogan, M. P. (2011). Development and evaluation of a staff training program on palliative care for persons with intellectual and developmental disabilities. *Journal of Policy & Practice in Intellectual Disabilities, 8*, 42–52.

Hall, I. J., & Johnson-Turbes, A. (2015). Use of persuasive health messages framework in the development of a community-based mammography promotion campaign. *Cancer Causes & Control, 26*(5), 775–784. doi: 10.1007/s10552-015-0537-0

Hsu, L. L., Chang, W.H., & Hsieh, S. I. (2015). The effects of scenario-based simulation course training on nurses' communication competence and self-efficacy: A randomized controlled trial. *Journal of Professional Nursing, 31*, 37–49.

Jackson, C. (2018). The public mostly trusts science. So why are scientists worried? *Science.* doi: 10.1126/science.aat3580

Kerse, N., Buetow, S., Mainous, A. G., Young, G., Coster, G., & Arroll, B. (2004). Physician-patient relationship and medication compliance: A primary care investigation. *Annals of Family Medicine, 2*, 455–461. doi: 10.1370/afm.139

Kolb, D. (1984). *Experiential Learning: experience as the source of learning and development.* Englewood, NJ: Prentice Hall.

Koponen, J., Pyorala, E., & Isotalus, P. (2012). Comparing three experiential learning methods and their effect on medical students' attitudes to learning communication skills. *Medical Teacher, 34*, 198–207.

Kreps, G. L. (2014). Evaluating health communication programs to enhance health care and health promotion. *Journal of Health Communication, 19*, 1449–1459. doi: 10.1080/10810730.2014.954080

Kreps, G. L., & Thornton, B. C. (1992). *Heath communication theory and practice.* Prospect Heights, IL: Waveland Press.

Laidsaar-Powell, R., Butow, P., Boyle, F., & Juraskova, I. (2018). Managing challenging interactions with family caregivers in the cancer setting: Guidelines for clinicians. *Patient Education and Counseling, 101*, 983–994. doi: 10.1016/j.pec.2018.01.020

Layton, D., Jenkins, E., McGill, S., & Davey, A. (1993). *Inarticulate science? Perspectives on the public understanding of science and some implications for science education.* East Yorkshire: Studies in Education.

Lipsey, M. W. (1993). Theory as method: Small theories of treatments. *New Directions for Program Evaluation, 57*, 5–38.

Makoul, G. (2001). Essential elements of communication in medical encounters: The kalamazoo consensus statement. *Academic Medicine, 76*, 390–393.

Mervis, J. (2017). Data check: U.S. government share of basic research funding falls below 50%. *Science* doi: 10.1126/science.aal0890

National Academy of Medicine. (2018). About the NAM. Retrieved from https://nam.edu/about-the-nam

National Academies of Sciences, Engineering, and Medicine. (2017). *Communicating Science Effectively: A Research Agenda.* Washington, DC: The National Academies Press. https://doi.org/10.17226/23674

Neuhauser, L., & Paul, K. (2011). Readability, comprehension and usability. In B. Fischhoff, N. T. Brewer & J. S. Downs (Eds.), *Communicating risks and benefits: An evidence-based user's guide* (pp. 129–148). Silver Spring, MD: U.S. Department of Health and Human Services.

Norgaard, B., Ammentorp, J., Kyvik, K., & Kofoed, P. (2012). Communication skills training increases self-efficacy of health care professionals. *Journal of Continuing Education in the Health Professions, 32*, 90–97.

Pew Research Center. (2017). The Partisan Divide on Political Values Grows Even Wider. Retrieved from: www.people-press.org/2017/10/05/the-partisan-divide-on-political-values-grows-even-wider/

Rajput, A. S. D. (2017). Science communication as an academic discipline: An Indian perspective. *Current Science, 113*, 2262–2267.

Rogers, C.L. (2000). Making the audience a key participant in the science communication process. *Science and Engineering Ethics, 6*, 553–557.

Ross, L. (2012). Interpersonal skills education for undergraduate nurses and paramedics. *Journal of Paramedic Practice, 4*, 655–661.

Roter, D. L., Wexler, R., Naragon, P., Forrest, B., Dees, J., Almodovar, A., & Wood, J. (2012). The impact of patient and physician computer mediated communication skill training on reported communication and patient satisfaction. *Patient Education and Counseling, 88*, 406–413.

Russell, S., Daly, J., Hughes, E., & op't Hoog, C. (2003). Nurses and 'difficult' patients: Negotiating non-compliance. *Journal of Advanced Nursing, 43*, 281–287. doi: 10.1046/j.1365-2648.2003.02711.x

Scheufele, D. A. (2006). Messages and heuristics: How audiences form attitudes about emerging technologies. In J. Turney (Ed.), *Engaging science: Thoughts, deeds, analysis and action* (pp. 20–25). London: The Wellcome Trust.

Sidani, S., & Sechrest, L. (1999). Putting theory into operation. *American Journal of Evaluation, 20*, 227–238.

Stewart, M., Brown, J. B., Weston, W. W., McWhinney, I. R., McWilliam, C. L., & Freeman, T. R. (1995). *Patient-centered medicine: Transforming the Clinical Method.* London: Sage.

Storksdieck, M., Stein, J. K., and Dancu, T. (2006). Summative evaluation of public engagement in current health science at the current science and technology center, Museum of Science, Boston. Annapolis, MD: Institute for Learning Innovation. Retrieved from: http://informalscience.org/images/evaluation/report_224.pdf

Stubblefield, C. (1997). Persuasive communication: Marketing health promotion. *Nursing Outlook, 45*, 173–177. doi: 10.1016/S0029-6554(97)90024-5

Sturgin, P., & Allum, N. (2004). Science in society: Re-evaluating the deficit model of public attitudes. *Public Understanding of Science, 13*, 55–74.

Sudore, R. L. & Schillinger, D. (2009). Interventions to improve care for patients with limited health literacy. *Journal of Clinical Outcomes Management, 16*, 20–29.

Train, T. L., & Miyamoto, Y. J. (2017). Encouraging science communication in an undergraduate curriculum improves students' perceptions and confidence. *Journal of College Science Teaching, 46*, 76–83.

Trench, B., & Miller, S. (2012). Policies and practices in supporting scientists' public communication through training. *Science and Public Policy, 39*, 722–731.

Vosoughi, S., Roy, D., & Aral, S. (2018). The spread of true and false news online. *Science, 359*, 1146–1151. 10.1126/science.aap9559

Wittenberg-Lyles, E., Goldsmith, J., Ferrell, B., & Burchett, M. (2014). Assessment of an interprofessional online curriculum for palliative care communication training. *Journal of Palliative Medicine 17*, 400–406.

Wynne, B. (1991). Knowledges in context. *Science, Technology and Human Values, 16*, 111–121.

Yuan, S., Oshita, T., Abi Ghannam, N., Dudo, A., Besley, J. C., & Hyeseung, E. (2017). Two-way communication between scientists and the public: A view from science communication trainers in North America. *International Journal of Science Education, 7*, 341–355.

Ziman, J. (1991). Public Understanding of Science. *Science, Technology, & Human Values*, *16*(1), 99–105. https://doi.org/10.1177/016224399101600106

10 A Metro for science communication

Building effective infrastructure to support scientists' public engagement

Brooke Smith

"Doors closing. Please stand clear of the doors." For anyone who lives, or has spent time in Washington, D.C., you recognize this as the announcement just before the Metro pulls away from a station. Public transportation systems, whether Washington D.C.'s Metro, New York's subway system or London's Underground, are designed to move a lot of people as efficiently, effectively, and safely as possible, to multiple destinations. While trains may not always run on time, and tracks may be under repair, not having this public transportation infrastructure would result in chaos during commuting hours, congested roadways and more. The Metro provides something critical to a buzzing, busy city – people-moving infrastructure, and it helps get more people to places efficiently and effectively. Without thoughtful transportation infrastructure, people expend more time and more resources to get to their destination.

As more scientists are interested in communication and engagement – the field of science communication research and practice is blossoming, more training programs are coming on line, and more diverse goals and objectives for communication are being articulated – it may be time to think about building infrastructure, or building a Metro, to support scientists' communication. [For the purposes of this chapter, the author defines "Communication" broadly (including definitions typically associated with communication, engagement and outreach) as the number of ways science (people, process, and content) connect with people.] Building an infrastructure for scientists' communication journeys has the potential to support more scientists to communicate effectively and efficiently, clearly articulate the diverse goals and objectives for communication, and create a system of practice based on robust research that is maintained through regular evaluation. At a time when scientists are expending a lot of time, energy and resources to communicate, and perhaps are not communicating as effectively as they could be, we should consider building an infrastructure to help move scientists to their communication destinations as effectively and efficiently as possible.

What does building and maintaining infrastructure entail? It takes a complex network, structural system, and multiple efforts to ensure the Metro can move people to multiple destinations every day (there are about 800,000 trips taken daily on D.C.'s Metro system). Infrastructures require not just "hard structures" (physical networks), but also "soft structures" (institutions and cultures). To build a robust infrastructure for a public transportation system, and arguably for science communication and engagement, you need the following: clear destinations, multiple pathways or lines, a transit authority, research (or engineering), policies, culture, and funding. An examination of each of these characteristics as they relate to an infrastructure for scientists' communication, is explored in this chapter.[1]

Metro stops: science communication destinations

There are multiple destinations on each Metro line. Rarely do riders board the Metro without knowing what their final destination is. Riders like to take the most direct way to their final stops. Yet, it's unclear whether scientists, or even those supporting communication and engagement (including practitioners and funders), are clearly articulating what their communication destinations are. Without these clearly articulated goals and objectives, scientists may be meandering about, aimless or misdirected in their communication efforts. If we don't know where scientists are trying to get to, we aren't able to effectively monitor and measure whether they've arrived.

There is a myriad of goals, or destinations, for science communication. A recent study by the US National Academy of Sciences named five distinct goals for science communication: to make the public aware of science, to build their appreciation of it, to develop their understanding of it, to inform their actions with it, and, in turn, to have science itself become more informed by the public (National Academies of Sciences, Engineering, and Medicine, 2017). A recent review of a history of science communication in the United States reveals that different social forces have resulted in the rise of different science communication goals and practices in science communication. Since the late 1800s, historical events from urbanization to the cold war to declining formal education structures, have influenced a series of communication goals for science including enlightenment, acculturation, demonstrating utility (and making a case for funding), supplementing the formal education system, empowerment and actualization (Bevan & Smith, forthcoming). Today, against a backdrop of misinformation and a rapidly changing media environment, scientist communicators are also looking for ways to ensure critical thinking is instilled in how we evaluate information and problems. Science communication goals may have emerged reactively to these social pressures. The consequence of this evolution may be that our current infrastructure's destinations were built reactively as these science communication goals evolved through history.

We do know that one prevalent goal of a scientist communicator is to rectify what they may see as a lack of knowledge in people with whom they are communicating. This deficit model approach to communication is not only proven to be ineffective (Kahan et al., 2012; Sturgis & Allum, 2004) it is often one of the most common goals held by scientists (e.g., Dudo & Besley, 2016). If "tell people information they should know, so they use it" is a destination on the

Metro of science communication, we may need to rethink how many scientists (if any) should have a ticket to arrive at that destination. A recent landscape study of North American science communication trainers suggests that communication trainers themselves may be enabling scientists' journey to this deficit model destination (Besley & Dudo, 2018). The authors noted

> even as many trainers may say they do not want to reinforce a deficit model of communication, many trainers seem to continue to emphasize that their objective is fostering a more informed public and that doing so will lead to better personal and individual decision-making.

This, and other research, highlights the need for communication trainers to help scientists explore specific communication objectives for particular contexts and audiences (Dudo & Besley, 2016) as well as the specific learning goals for communication training (Baram-Tsabari & Lewenstein, 2017).

Scientists, and those supporting scientists to communicate, should identify these destinations. "Scientists' most important communication decision may be figuring out their goals. Do they want to help shape local, state or national policy discussions? Do they want to influence individual behavior, such as diet choices, medical decisions or career paths?" (Dudo & Besley, 2016).

Having consensus on what the destinations even are, and alignment on what we define as a goal versus an objective, will be key to building an infrastructure. Ensuring that we are inclusive in how we consider defining these destinations, so that it reflects ideas from a truly diverse set of scientists, communities and people, will be imperative in ensuring the entire infrastructure is resilient, useful, and lasting. And, ultimately, ensuring that the destinations we have identified are not only where scientists want to go but also where people are and want scientists to journey to, will be important.

Metro lines: pathways for communication and engagement

Getting to a destination requires a specific line, or pathway. If you are trying to get to Ronald Reagan Airport, you have a couple of lines to choose from. You could get there on either the blue or the yellow line. But you certainly wouldn't take the red line towards Silver Spring. In science communication, finding the right line or pathway can help scientists get to their destination. How many scientists may be on a red line, when a yellow line stop is really their destination? If scientists are able to articulate their communication destination and goals, the next question is "how do they get there?"

There are a remarkable number of pathways available to scientists, to help them navigate their way to their communication destination. Facilitators of engagement and communication live at a myriad of institutional homes including universities, museums, media outlets, scientific societies, community organizations, think tanks, private companies, policy or political institutions and more. These facilitators of engagement often have connections in the world of science, but also to the audiences relevant to the goals that scientists seek. For example, if a scientist's goal is to "inform policy," that scientist may wish to work with their disciplinary society's government affairs office, with an outside boundary organization (such as Pew's Lenfest program for ocean scientists) who can help connect scientists with relevant public policy and political audiences, or with a local civic group familiar with local/state politics. If a scientist wishes to share their ideas as a public influencer, they could work with their press information office to connect with journalists, or they could write and place their own opinion pieces in media outlets or write for an outlet such as *The Conversation.* If a scientist is interested in participating in the education process, with the goal of enhancing children's science education, they might connect with their local school district or reach out to their local science center or museum to interact with children in more informal settings.

The field of facilitators of communication is ever growing, and increasingly more connected. Within the informal science education system, experts to help navigate these pathways connect together through the National Informal Science Education Network (NISENet) or through the Association of Science and Technology Centers (ASTC). The Science Festival Alliance works to connect all local science festivals across the country. Engineers and Scientists Engaging in Policy (ESEP) is a network of scientists and professionals working to ensure science and policy are connected. The National Alliance for Broader Impacts is a national network of people working to build institutional capacity for the broader impact of science. The National Academy of Science's Entertainment Exchange brokers connections between scientists and entertainment industry professionals. There is an effort afoot to better connect these existing networks, further bringing structure to the pathways and lines available to scientists.

A recent landscape overview of facilitators servicing different goals revealed that they have a number of things in common. The majority of them work to humanize scientists, they work to connect as diverse a group of scientists as possible, and the audiences they serve want to interact directly with scientists. Facilitators also commented how scientists they work with become more skilled at communication and what it entails, the more they engage. And finally, very few facilitators are current, or stay current, on the literature about informal education or science communication. (Gentleman, Weiner, Cavalier, & Bennett, 2018)

While there are a number of organizations, people, and pathways available to help scientists connect to their communication goals, not all of them are familiar to scientists. Conversations on this topic have revealed that scientists don't know the full range of pathways available to them. As a result, scientists might find themselves on a pathway recommended by a peer, or connect to a local school (especially if they have relatives there). While these are legitimate pathways, they may not be the ones best suited to help a particular scientist journey to their destination.

Furthermore, a critical lens should be brought to look at who are providing pathways, and who is missing from that group. Ensuring pathways, and those supporting them, reflect the diversity of our population (not just the dominant gender or race in science itself), will also be imperative in ensuring a sustainable and effective infrastructure.

Communication trainers can help scientists develop communication skills, but are they linking scientists to pathways to do meaningful engagement? Are these pathways clear even to trainers? Should additional pathways be sought out? Often communication trainers help scientists to "practice," but are they putting scientists in enough "games"? (Besley & Dudo, 2018). An infrastructure can better link practice in trainings, to performance in real world communication opportunities.

Engineering the Metro: using research and data to build and evaluate science communication

Designing the Metro took considerable research, design and engineering. Thorough research about how to work with existing neighborhoods, geologic structures, traffic patterns, and more were all necessary to ensure the Metro was built on our best available understanding of the landscape, geography, demographics and more. Since the Metro was built, it has been monitored. How many people are using each of the stations? Is there significant use or congestion on certain lines on certain times of day? Is the physical integrity of the tracks intact? Designing our science communication pathways and monitoring our destinations (have people actually arrived there? are people even using them?) based on research and data should be foundational as we build and practice science communication. However, ironically, those communicating about science rely more on intuition rather than scientific inquiry (Kahan, Scheufele & Jamieson, 2017)

There is a wealth of research about science communication, public engagement with science, and informal science learning. While there is a science of science communication field, many research fields feed into this field or are relevant to the relationship between science and society. Communications, education, risk, psychology, neuroscience, sociology, behavioral science, decision-theory, political science, and more, all contribute to our understanding of how science and society interact and relate to one another. Bringing knowledge and insights from these fields to the practice of science communication is imperative.

Science about science communication is not always, but is increasingly, used in practice. People who facilitate science communication pathways have noted that, for the most part, they don't actively seek out social science to inform their practice (Gentleman et al., 2018). Science communication trainers have had a mixed view about the value of science of science communication research – they are able to name a few researchers and studies, but also note the research is rather inaccessible (Besley & Dudo, 2018). Some fields, like informal science education, are perhaps further along in a commitment to connect research to practice.

Researchers are working to make their theory, knowledge and insights more accessible to practitioners. For example, the National Academy of Science's

Sackler Colloquia on science communication is intended to have social science researchers share their knowledge with practitioners. The Center for the Advancement of Informal Science Education (CAISE) curates a website of science, research and other resources easily accessible to those in the field.

In addition to basing engagement opportunities or training programs on research, there is also a need to collect data and information about whether engagement efforts or communication trainings are effective at achieving a defined goal. Evaluation in the informal science learning space has been advancing. The training community rarely collects information about scientists' communication progress. Often when they do collect information, it is about scientists' views of the training event and/or their own perception of their efficacy (Besley & Dudo, 2018). However, this trend is shifting. For example, Rodgers et al. (2018) put forward an innovative framework to assess the effectiveness of their science communication training program.

We must build our practices and infrastructure based on what we know, and then observe, learn and iterate, to ensure our infrastructure is effective at helping scientists reach their goals, and to reevaluate the value of current destinations and whether new stops are needed.

The Metro Transit Authority: central coordination for science communication

The Washington Metropolitan Area Transit Authority is a multi-jurisdictional government agency that serves as the lead Metro authority. Many other institutions are involved (e.g., Metropolitan (D.C.) Police Department, Fairfax County, the State of Maryland, etc.). For the Metro to work, multiple players coordinate, constantly.

Disparate science communication efforts are largely uncoordinated, leading to what some have described as a largely fragmented science engagement ecosystem (Lewenstein, 2001). Multiple organizations play a role in scientists' communication, including universities, industry, NGOs, government agencies, funding agencies, boundary organizations, the National Academies, and scientific societies. We do not lack for organizations that are contributing and wanting to contribute more. We do not have, nor has the community fully explored, a science communication "transit authority" – a thoughtfully guided central entity or coordination mechanism for the various institutions who play a role in supporting scientists' communication journey. A central entity – formal or informal – could help scientists navigate their way through the infrastructure, helping them identify their goals and appropriate training and engagement pathway partners, to pursue those goals.

On many university campuses alone, there can be multiple programs and people who support scientists' communication and engagement. They don't always know about each other or talk to one another. One department may develop their own communication training or course, while another department might (simultaneously) bring in an outside trainer to work with their faculty.

Similarly, one department may have an outreach program to local schools, while another research project receives funding which includes support to local school outreach. On some campuses, there are central offices to help coordinate across different engagement goals and opportunities. For example, The Connector at University of Missouri acts as a hub for professional development, engagement opportunities, and measuring impact – bringing together professional staff, faculty, community members and expert resources to achieve communication and engagement opportunities, geared at specific goals. Far too often, the full suite of resources available to scientists on a single campus is invisible and not connected up together.

In countries outside of the United States, there are efforts to coordinate (if loosely) science communication training and practice. In Europe, governments, higher education institutions, research councils and the European Commission all play roles in supporting scientists' communication. These roles can be contradictory, in some instances, and these institutions may not distinguish between training scientists and capacity building of science communicators (Trench & Miller, 2012).

The increase in coordination of networks in this space – among groups like the National Alliance to Broader Impacts, the Association of Science and Technology Centers, science communication trainers, Center for the Advancement of Information Science Education and more – may provide opportunities to create more centralized on-ramps, off-ramps and traffic control for scientists' activities. Ensuring these groups, and opportunities, are known, visible, supported and accessible will be critical in maintaining a structure.

The Metro's rules and social norms: supporting policies and a culture for science communication

There are policies, such as "no food or drink on the Metro" or "no standing on the edge of the tracks" to be sure Metro rides and riders are safe. There is also a strong Metro culture, how people behave and treat each other while all trying to efficiently get to their destinations. The Metro's escalator norm to "stand on the right, walk on the left" is so strong that people have been chastised for standing on the left. Other cultural courtesies exist, like letting everyone off at a stop before people get on. All of these policies, cultures and norms support certain expected behaviors in order to help reduce congestion, keep people safe, and get the greatest number of people to their destinations as efficiently as possible.

An exploration of science communication norms, policies, and culture deserves a critical look as we build the science communication infrastructure. The norms and policies of the Metro enable a more efficient and effective system. The norms and culture in science and science communication should enable a system that moves scientists to communication destinations efficiently and effectively. However, our norms, cultures and policies in science communication may not be just unsupportive of scientists' communication, they may be obstructive.

Certainly, the elephant in the room when discussing scientists' ability to communicate and engage is the lack of institutional support, recognition, career advancement and rewards for doing so. Currently, the culture of science does not support a culture of communication engagement. Research funding and publications are the coin of the academic realm. Supporters and practitioners of public engagement consistently face challenges about the lack of recognition for this work in promotion and tenure decisions. While there is a call (and perhaps an increasing one) to improve the support and recognition of scientists' communication, implementing these values into a scientific reward system has been challenging. We lack clear measurement and guidelines by which to evaluate scientists' communication efforts. Often, academic leadership speaks to the importance of communication and engagement, but this is often merely lip service without real policies and resources trickling down to departments managing the work (Risien & Nilson, 2018).

There is hopeful anticipation that the culture may be changing as young scientists, especially graduate students, demand communication training in their graduate education and look to find ways to include communication as a central part of their career. There has been a call for Graduate STEM education to teach and prioritize communication (Neeley, Goldman, Smith, Baron, & Sunu, 2014). Efforts like ComSciCon, a group of graduate students that has created a communication training program by and for graduate students, underscore this shifting landscape. Risien and Nilson uncover two flaws in this narrative of hope for the future generation of scientists to create a culture of engagement: 1) attrition, those who really value engagement are unlikely to stay in the academy, and 2) those who do value it and stay in the academy are likely to downgrade their level of engagement to fit within disciplinary norms and improve job security (2018).

In addition to these cultural norms not supporting engagement, other "policies" for science communication need to be considered as the infrastructure is built. Some have pointed to a need to discuss and explore the ethics of science communication, and the ethics of various tactics and skills in science communication. It has been argued that it is time we consider what an ethics of science communication might look like – exploring what really makes communicating science a good, moral thing to do (Medvecky & Leach, 2017). As Priest, Goodwin and Dahlstrom put it:

> I have met quite a number of scientists who see science communication as a strategy for getting others 'on their side' and yet at the same time they seem to feel that the information they have to offer is (or should be) entirely neutral. This is a deep tension in the field of science communication, but one that is not commonly recognized or discussed; neither scholars nor practitioners commonly question whether it is possible to do both – nor how often one sometimes masquerades as the other.
>
> (Priest, Goodwin, & Dahlstrom, 2018)

Getting past goals and strategies, what are the ethical considerations for various tactics of communication? As storytelling and narrative become an increasingly popular tactic in science communication training (Besley & Dudo, 2018), we might also be asking ourselves to evaluate the ethical considerations. What are the underlying purposes of narrative: comprehension of persuasion? What are the appropriate levels of accuracy to maintain? Should narrative be used at all? (Dahlstrom & Ho, 2012)

If scientists enter an infrastructure with an ability to define a clear destination, identify a clear pathway to get there, and bring research and evaluation to their work, will the infrastructure truly be able to support them if there are cultural and policy norms that have not been defined or are working against them?

Financing and building the Metro: investing in the science communication infrastructure

Most public transportation systems rely on a number of different sources of revenue, including fares, advertising, and local government subsidies. The Metro's agency funding comes from a number of government agencies including the District of Columbia, Prince George's County, Montgomery County, Fairfax County, Arlington County and the cities of Alexandria, Falls Church and Fairfax (Washington Metropolitan Area Transit Authority, Approved Fiscal Year 2016 Budget). There is revenue generated from those who use the Metro, as riders pay a fee to access the infrastructure to get them to their destination, and of course there are ads throughout the Metro station and on the trains themselves that bring revenue to the transit system.

How we pay for building and maintaining the science communication infrastructure is critically important, but drastically under-discussed. In a 2014 discussion at the National Academy of Sciences about infrastructure, a panel of experts discussed the US investment in this scientific infrastructure, and explored what we currently invest in science communication. It turns out it is extremely difficult to identify a specific number for this – largely because science communication funding is usually not its own line item in a budget one can aggregate. Given what we do know about funding in science from government and philanthropy, we can speculate that it is a very small percentage of what we invest in science research (perhaps as low as 0.1%, although 1% is more likely; the figure is unlikely to be over 5%). This same discussion, though, noted that without clear destinations defined, return on investment cannot be described, and therefore investment won't flow (PILS sustainable infrastructure report, National Research Council, 2014).

A lot of key questions exist as we consider how the building and maintenance of an infrastructure should be financed, and by whom. Should scientific research funding allocate a percentage of its funding (overall or per project) to communicating science? Should the federal government support this? Should scientists using the infrastructure pay, like a rider does, for its support? Perhaps scientists should pay to use it, but not pay to build it? Is this infrastructure a public good and something the public is willing to invest in?

Sustainable financing for any kind of infrastructure (public transportation, water, roads) is always a challenge. People expect great infrastructure to simply exist, but they often don't want to pay for it – or wait for it to be developed comprehensively. As a result, infrastructure often fails, and needs to be repaired instead of maintained, which can be even more costly. Is the same true for science communication?

The role of the science communication training community in building, maintaining, and moving scientists through the science communication infrastructure

What role and responsibility do those who practice or research science communication training have in building, maintaining and being a player in this infrastructure? While there is a need for thoughtful leadership, vision and coordination across the entire infrastructure, those working on science communication training might ask themselves how to contribute to a few of these key infrastructure components:

Destinations and goals: Are training programs helping scientists identify their communication goals or objectives? If the program is tailored to support scientists in achieving a specific goal or objective (e.g., hands on learning about scientific process or informing policymakers), are they clear about their niche to scientists they support? Are trainers connecting the ideas of goals and objectives to the skills they teach? Are the goals and objectives scientists are considering representative and inclusive?

Pathways and lines: Are training programs connecting their practice sessions to real world engagement opportunities?

Research and evaluation: Are training programs using research about science communication to inform their curriculum? Are programs collecting data about impact and evaluation to improve or tweak their practices?

Building and maintaining infrastructure as it relates to creating science communication policies, changing culture, managing a central authority, and financing a Metro may seem bigger in scope than training practitioners and researchers can take on individually. But being an active voice and participant in field-wide conversations about these efforts and trainer/researcher/user experiences is critical.

Goals of science communication should be inclusive. All scientists should have access to this infrastructure, and this infrastructure should be built by a diverse group of people – to ensure it is resilient and accessible to all.

Looking ahead: envisioning a Metro for science communication

When the Metro is not running, those who are most determined to get to their destination find a way. They walk, or they pay extra for a taxi or rideshare. It takes them more time or more resources, but they do it. Similarly, in science communication today, determined scientists who aren't supported by an efficient infrastructure are in fact finding ways to reach their destination. It simply takes them more time and more resources.

Imagine if we had a functional infrastructure for science communication and engagement. More scientists could move to different communication destinations. It might be easier for those already engaging to do it more efficiently and effectively. We could support exponentially more scientists to engage and communicate with multiple audiences for multiple reasons. We could better understand the various goals, value of different pathways, and potential return on investment. We could learn from what was working and discuss what was missing. At a time when scientists are showing great appetite to engage, and at a time where the public's trust in science remains strong but their connection to science tenuous, it is time to build an infrastructure that channels scientists' passion and energy, while simultaneously ensuring communication efforts are efficient, effective (with measurable impact), sustained, and supported, both culturally and financially.

Note

1 This chapter is based on my synthesis comments for the National Academy of Sciences workshop, "Sustainable Infrastructures for Life Science Communication". The ideas were summarized on a COMPASS blog post (www.compassscicomm.org/blog/building-a-metro-for-science-communication). I would like to acknowledge the National Academy of Sciences workshop that birthed these ideas, and the COMPASS blog for being the first place the concept was initially described.

References

Baram-Tsabari, A. & Lewenstein, B. V. (2017). Science communication training: what are we trying to teach?. *International Journal of Science Education, Part B*, *7*(3), 285–300.

Bevan, B. and Smith, B. (forthcoming). History of Science Communication in the US: It's complicated. In: T. Gascoigne (Ed.), *The emergence of modern science communication.* Australian National University Press.

Besley, J. and Dudo, A. (2017). *Landscaping Overview of the North American Science Communication Training Community.* Available at http://informalscience.org/sites/default/files/Communication%20Training%20Landscape%20Overview%20Final.pdf (accessed February 6, 2019).

Dahlstrom, M. F. and Ho, S. S. (2012). 'Ethical Considerations of Using Narrative to Communicate Science'. *Science Communication 34*(5), pp. 592–617. doi: 10.1177/1075547012454597

Dudo A., & Besley J. C. (2016). Scientists' Prioritization of Communication Objectives for Public Engagement. *PLoS ONE 11*(2): e0148867. https://doi.org/10.1371/journal.pone.0148867

Gentleman, D., Weiner, S., Cavalier, D., & Bennett, I. (2018). *Landscaping Overview of U.S. Facilitators of Scientists' Engagement Community.* Available at http://informalscience.org/support-systems-scientists'-communication-and-engagement-workshop-iv-landscaping-overview-us (accessed February 6, 2019).

Kahan, D. M., Peters, E., Wittlin, M., Slovic, P., Ouellette, L. L., Braman, D., & Mandel, G. (2012). The polarizing impact of science literacy and numeracy on perceived climate change risks. *Nature climate change*, *2*(10), 732.

Kahan, D. M, Scheufele, D., & Jamieson, K. (2017). Introduction: Why Science Communication. In K. Jamieson, D. A. Scheufele & D. M. Kahan (Eds.), *The Oxford Handbook of the Science of Science Communication*. Oxford, UK: Oxford University Press.

Lewenstein, B. V. (2001). Who produces science information for the public?. In J. Falk, E. Donovan & R. Woods (Eds.), *Free-Choice Science Education: How We Learn Science Outside of Schools* (pp. 21–43). New York: Teachers College Press.

Medvecky, F. and Leach, J. (2017). The ethics of science communication. *JCOM, 16*(04), E.

National Academies of Sciences, Engineering, and Medicine. (2017). *Communicating Science Effectively: A Research Agenda*. Washington, DC: The National Academies Press, 17–20. https://doi.org/10.17226/23674.

National Research Council. (2014). *Sustainable Infrastructures for Life Science Communication: Workshop Summary*. Washington, DC: The National Academies Press. https://doi.org/10.17226/18728.

Neeley, L., Goldman, E., Smith, B., Baron, N., & Sunu, S. (2014). GradSciComm Report and Recommendations: Mapping the Pathways to Integrate Science Communication Training into STEM Graduate Education Available at www.informalscience.org/gradscicomm-report-and-recommendations-mapping-pathways-integrate-science-communication-training (accessed February 6, 2019).

Priest, S., Goodwin, J., & Dahlstrom, M., (Eds.) (2018). *Ethics and practice in science communication*. Chicago, IL, U.S.A.: University of Chicago Press. https://doi.org/10.7208/chicago/9780226497952.001.0001.

Risien, J., & Nilson, R. (2018). *Landscape Overview of University Systems and People Supporting Scientists in their Public Engagement Efforts*. Available at http://informalscience.org/sites/default/files/University%20SystemsPeople%20Landscape%20Overview_Risien_Nilson_March_2018_to%20Post.pdf. (accessed February 6, 2019).

Rodgers, S., Wang, Z., Maras, M. A., Burgoyne, S., Balakrishnan, B., Stemmle, J., & Schultz, J. C. (2018). Decoding Science: Development and Evaluation of a Science Communication Training Program Using a Triangulated Framework. *Science Communication, 40*(1), 3–32.

Sturgis, P., & Allum, N. (2004). Science in society: re-evaluating the deficit model of public attitudes. *Public understanding of science*, *13*(1), 55–74.

Trench, B., & Miller, S. (2012). Policies and practices in supporting scientists' public communication through training. *Science and Public Policy*, *39*(6), 722–731.

Index

Page numbers in **bold** denote tables, those in *italics* denote figures.